Penguin Education

Systems Analysis

Edited by Stanford L. Optner

Penguin Modern Management Readings

General Editor

D. S. Pugh

Systems Analysis

Selected Readings

Edited by Stanford L. Optner

Penguin Books

Penguin Books Ltd, Harmondsworth,
Middlesex, England
Penguin Books Inc., 7110 Ambassador Road,
Baltimore, Md 21207, USA
Penguin Books, Australia Ltd,
Ringwood, Victoria, Australia

First published 1973
This selection copyright © Stanford L. Optner, 1973
Introduction and notes copyright © Stanford L. Optner, 1973
Copyright acknowledgement for items in this volume
will be found on page 331

Made and printed in Great Britain by
Richard Clay (The Chaucer Press) Ltd, Bungay, Suffolk
Set in Monotype Times

To George and Ruth Lee

Contents

Introduction

For a time, the national pastime of professional detractors in the USA was to attack the dehumanizing effects of the *computer*. The *giant brain*, or the *monster* as it was lovably called, reigned for fifteen years as the undisputed butt of practical jokes about man's submission to the machine. But now the detractors will have to find a new target; witness this article taken from the *Los Angeles Times*, 1972, titled 'A giant step backward . . . humans replace computer in teacher credential office'[1]:

Sacramento (UPI) – George Gustafson has taken a giant step backward. He fired a computer. Gustafson, executive secretary of the State Teacher Preparation and Licensing Commission, fired the machine and switched to human beings. The move cut teacher credential processing time by 900 per cent.

Gustafson, Monday, described his demodernization program as 'converting credential processing from a complex and costly automated system to a streamlined, fully manual operation.' In other words, Gustafson said, 'It was easier to do it by hand. We got rid of a million and a half dollars worth of computer. We pulled the plug and sent it back to IBM.'

By taking the step, Gustafson said, he was also able to reduce the staff from 240 persons to 106. At the same time, getting rid of the new computer cut the credential processing period from an average of 95 days down to the current average of 10. Since August, the department has cut 80 employees. Gustafson said all but three were given jobs with other State agencies. A side benefit of the switch came during the summer, when the commission hired more than 50 poor and minority students to help make the change to manual processing.

Although the computer was unable to hold its job, Gustafson is the last person to pick on the unemployed machine. He said it was a 'good worker, it just couldn't compete with people.'

With the argument against the power of the computer somewhat vitiated, perhaps the new target will be *systems analysis*.

For a long time, I had the suspicion that systems analysis was a kind of technical status symbol of the great industrial nations. The appearance of the Nikoranov material (see Reading 8) told me that even the USSR as a matter of prestige wanted to be in on the act. I also noted that the Russians did not claim to invent

1. Tuesday, 18 January 1972, Section 1, p. 3.

systems analysis, something which has added doubt to my view of its ultimate future.

In the period after the Second World War, the United States inherited a legacy of operations analysis material from Great Britain. Ever since, there has been a great deal written about the role of systems analysis, and its use *has* placed the USA in the forefront of problem solving with computer technology. Perhaps there is some relationship between the number of successful users of systems analysis, and the state of technical problem solving in those firms. And perhaps systems analysis has contributed to the notion that we owe some of our economic success to its use.

Yet in Japan, where the number of computers in use is second only to the United States, I was unable to find a representative paper for this book. It also has occurred to me that the Japanese, who have brought systems design to a high state of development in banking (for example), may somehow practice systems analysis without using our western process. I understand, from the director of a research institute in Tokyo, that Japan wants to be a leader in satellite technology. If I can believe Thome and Willard (Reading 12), it can hardly be done without an intensive application of systems analysis. Therefore, I must conclude that systems analysis will either be used in its uniquely Japanese form, or it will be borrowed from the British, American or French model, whichever one may be used.

Can one practice systems analysis without knowing it? The answer is a vehement *yes*. Systems analysis should be viewed as the most recent and perhaps most comprehensive vehicle for complex problem solving. It is clearly the *result* of a long history of problem analysis using the heuristic (learn and relearn) technique. It is not completely new or different; it is not unique. Indeed, its power (if any) resides in the way its method synthesizes the method and the approach of great experimenters of history. As Hitch points out in Reading 1, systems analysis borrows from every scientific era, even from the Greeks of the fifth century BC.

Business and industry have emerged as the societal groups making the most use of systems analysis. But not far behind are the business applications of Federal, State and local government (see Reading 14). The complex management problems of government, and the availability of the computer to attack them (notwithstanding the California Teacher Credential Office), have spawned new versions of old techniques, such as PPBS (Planned Programming Budgeting System). Since systems analysis is a kind of midwife to all computer applications, it will doubtless be used and misused in an increasing variety of ways.

Sad to say, the use of systems analysis is no guarantee in itself that the user will derive an ideal solution to a problem. From the earliest period of its use, both theorists and practitioners have a concept of high quality solutions which can be derived by applying systems analysis. I question whether we analysts do enough in the way of designing the system to exploit the potential in the method. Perhaps this explains why so many solutions seem derivative, and not freshly wrought-based upon all the evidence and all the available technology. Perhaps the continuing inability to exploit this tool of the computer era explains why so many data processing systems are automated copies of the original manual systems.

I believe we are still learning how to engage with problems and to intervene with existing systems. To engage with a problem, the systems analyst must do as Hoag, Jordan, Quade and others suggest: predesign the study and the problem-solution process, and be prepared to evaluate every successive step in the systems analysis. Intrinsic to the ability to engage is the selection of objectives and constraints, the identification of alternatives and their component systems costs by which to evaluate the possible courses of action.

Intervention of the existing system takes place when the alternative has been selected and the system is in the period of installation. Intervention is the orderly process of replacement, when the existing process is phased-out, and the new system is phased-in.

In the case of the Teacher Credential Office, the reader may draw his own conclusions about the area of the problem: did the systems analysis fail to engage the problem? Or was the failure in the installation process? How many of us actually accept the fact that 'the computer just couldn't compete with people'? Not I for one! The failures may have occurred in either or both areas, but whatever the reason, it was not the fault of the witless machine to which human propensities have been attributed. In an ultimate way, all failures are those of people.

Systems analysis, in its earliest business applications in 1956, was influenced by industrial engineering. However, industrial engineering lacked precedents for combining a number of professional disciplines in a single project effort. The typical areas of industrial engineering (time and motion study, work methods, plant layout, systems and procedures) were aimed at manufacturing support and work measurement. Unit record (tabulating) equipment was slow and had limited capabilities for improving the effectiveness of procedures.

In business problem solving, the applications of operations research, management science and computer technology were not yet proven. When I was at the Ramo-Wooldridge Corporation (now TRW), from 1956–58, my department was engaged in consulting contracts in which the multidisciplinary approach was tried. But the effort was a failure. The mathematician with the most impressive academic laurels held the blackboard and the attention of management. Project personnel who knew the most about the client problems had almost no influence. In the course of project direction, scientists tried to fill the management role. But they had no training in business; to them, every problem was seen as a potential algorithm.

Just a few months before the Division of 300 technical staff was dismissed, the Ramo-Wooldridge management gave me (with my industrial engineering background) the opportunity to organize a systems approach. But it was too late for that organization. Today, it seems ironic that we were at the threshold of a new industry, but neither the technical staff nor the management recognized the competitive opportunity at hand. Companies like The RAND Corporation and Systems Development Corporation were completely dedicated to military problems at that time. Stanford Research Institute, A. D. Little and other technical consulting companies had relatively narrow capabilities.

It is noteworthy that Ramo-Wooldridge already had the necessary 'systems approach'. (Note that Simon Ramo, the author of the last Reading, still uses the term, as does Churchman in Reading 16.) Engineers and scientists were using the systems approach in a building barely 500 feet away from my office. They were applying the scientific method to solve problems of equipment design, missile tragectories and project control in the Atlas, Thor, Titan and Minuteman programs of the US Air Force. The work of the Guided Missiles Research Division (GMRD) was to integrate, monitor and support the twenty-odd prime contractors in the ballistic missile program. This cooperative research and development role of Ramo-Wooldridge was called systems management; the goal of the systems managers was to apply systems analysis in an effective, analytical, problem-solving effort.

The technical staff of GMRD was trying new ways to achieve their objectives, buttressed by the traditional research methods of scientific disciplines. In each discipline, there already were proven routines of analysis and synthesis, design and implementation. The integrating concept of the multidisciplinary effort was the system. The underlying assumption of that concept was that the disciplines interlock. The concept also required that each discipline implement its research and development in recognition of the total systems requirements of the end product, and not place disproportional emphasis upon any one subsystem.

But the members of the technical staff in the consulting, business-oriented division never gained access to GMRD because we had no 'need to know'. Thus, in 1958, although some of us had inklings of a systems approach, we were fenced off from the basic concepts of systems analysis which were yet to be articulated for business applications. As Nikoranov points out (Reading 8), the first books which were exclusively occupied with systems analysis only began to appear in 1960. By 1965, there were many books, and at least two societies (Operations Research Society of America and the Institute of Management Sciences) which devoted space in their publications to the components of systems analysis, their content and method.

Two factors moved systems analysis into the forefront as a discipline for the computer in the 1960s. The first was the need for excellence in applications of the new technology in thousands of different areas of business, government, the sciences and the academic world. Systems analysis mediated between the problem and the means employed to solve it. Some analysts applied it with more success than others.

The second factor influencing the development of systems analysis was the classroom process. It was essential that educators provide a format to their students, especially those who were aiming at careers in business or government as computer programmers or systems analysts. At the outset, the classic argument was whether or not to train professional specialists (traffic engineers, librarians, production control experts, etc.) to be programmers. The converse argument was whether or not to train programmers, to the extent necessary, in a professional specialty. The latter practice has been more widely used, and it is typical that a large company may have six or more programmer levels, and another six or more systems analyst levels. To begin an industrial or government career, the trade school or college-educated programmer starts as an apprentice, at a low place on the table of organization.

In Parts Two and Three of this book, the evolutionary process of systems analysis is captured in readings. My own experience, recounted above, is representative of many systems analysts. As consultants to government, we had frequent opportunities to break into new applications areas, despite our lack of professional qualifications for each particular area. In those days (1963-8), no consulting company had exceptional qualifications to carry out fully implemented data processing systems. Therefore, our way of exercising caution was to attack a new problem in the most methodical way possible. After a few learning experiences, the methodical systems approach became a prescriptive practice . . . the way to do it!

For example, the first principle we formulated was that the investigator should document, in the initial phase of the systems analysis, exactly how the client currently handled his problems. This led to the formulation of procedures we now call the existing systems study. The ideas for the requirements analysis, the development of systems requirements and the design of a proposed system were similarly evolved. These provided methodical ways to improve the quality of applications, and to reduce the risk of failure in some subsequent stage of development.

The detailed systems design, which follows the proposed system sequentially, is critical in the systems analysis. As late as 1971, my Japanese colleagues told me that they felt that systems design was their least satisfactory performance area; in America, it is certainly far from a fool-proof procedure, witness the article by Jaffe (Reading 13). The subsequent steps involving logical design, programming, testing, documentation, training, installation, parallel operations and production are all a part of what we call the implementation effort. They are also integral parts of a systems analysis.

Given this methodology, with its trial-and-error development over a period of fifteen years, the experience of the Teacher Credential Office takes on another aspect. Systems analysis has become the vehicle for handling the conversion to automatic data processing of very large-scale, complex problems. But, despite the availability of the method, if systems analysts do not deal completely with the application, significant problems can arise. For instance, in 1966 my consulting group processed 450,000 borrower cards for the Los Angeles Public Library. The cards consisted of the borrowers' full names, and the procedure went well except for one thing: for economic reasons we had truncated first names after the sixth letter. When the cards were produced, all the Virginias appeared as 'Virgin' and all the Christophers as 'Christ'! (My card read 'Stanfo'.) The library was inundated with phone calls. However, we solved the problem within a week by passing the name file against a computer program which expanded the common first names to the full lengths, and printed new cards.

This error was the oversight of thirty experienced library personnel and consultants; we all missed this human relations problem. We were all guilty of passing over critical review steps in the proposed systems design. Perhaps it would have been caught if a selective batch of borrower cards had been produced in advance of printing out and mailing the complete file.

In any event, it was only one problem, of greater concern to some than to others. Like the Teacher Credential Office, the computer became the butt of the joke, and only a few people saw beyond the alleged failure of the machine. Today we anchor our systems analysis in the knowledge that the machine is error-free, one constant in a sea of variables, most of which are within the control of the systems analyst. I predict that the Teacher Credential Office will one day be a computer user again when the number of people required to operate the manual system overflows the space available, or when the number of transactions and the cost per transaction outrun the State government's willingness to pay the price. Some day the Teacher Credential Office will discover systems analysis and leap into the age of penicillin, television and the computer.

I would like to thank Mrs Shirley Frost who assisted me in proofreading. Special thanks go to my wife, Ruth, who not only read and edited, but helped form the finished work. I would also like to take this opportunity to thank my contributors, their publishers and my editor at Penguin for making this book possible. And a positive word of thanks to my colleagues and clients who provide me with the continuing ammunition of systems analysis.

Part One
Origins and Development
of Basic Concepts

The initial Readings in this book reveal a heated discussion about systems analysis which was ongoing in the early 1950s. Professionals with training in the social and physical sciences and management specialists with experience in weapons evaluation and procurement were agreed: the programs of defence after the Second World War required a more far-reaching problem-solving technology. However, it was clear that there were mixed opinions on the efficacy of systems analysis in its problem-solving role.

The papers from The R A N D Corporation testify to the use of a new methodology heavily influenced by operations research and mathematics. The need for precision in solutions was overbalanced by an increasing appreciation of the complexity inherent in the simplest military problems. Thus, the ability to evaluate alternatives in a consistent unarbitrary manner emerged as a principal objective of systems analysis.

The writing of the 1950s and early 1960s illustrates a preoccupation with *uncertainty*, a new use of the term which was coined to identify the part of the problem least susceptible to precise solution. Given the impact of uncertainty upon a particular alternative, the authors describe how they attempted to render their problem solving meaningful. The idea that *numbers are solutions* may have been a way of dealing with uncertainty, and the use of probability estimate s can be justified on this basis. However, in parallel with the mathematical attack on problem solving is the common-sense approach which includes practical considerations, also.

The rising emphasis on *criteria selection* was a reflection of the practical considerations. The earliest systems analysts were aware that improperly selected criteria would overthrow the best technical solution. Perhaps it was the practical considerations of business and industry that gave rise to successful applications of systems analysis. As a result of dealing with practical problems, there were dramatic developments in the 1960s of new ideas associated with the use of systems analysis in aerospace and in industry.

The papers of the 1960s reflected a continuing effort to deal with the role of computers in problem solving. The introduction of new subjects such as *self-organizing systems, perception* and *cognition* in a systems context constituted a step away from exclusively military applications, and a step toward the use of systems analysis in a variety of social and industrial problems.

The use of systems concepts in research became more articulated in the mid-1960s. Classification, definition and other organizational attempts to bring the systems approach into research were underway. Wright, for example, discussed systems in cultural terms. He illustrated how the systems analyst could bring mathematical tools together with a model of the system for a military problem. He then proceeded to show the applicability of the same principles in a social context, drawing propositions and postulates.

Hall was among those who brought systems into the engineering of complex objects, defining some basic terms and describing the environment of systems. He was concerned about the properties of systems, and discussed *natural* and *man-made systems*. He also described the property of *feedback in ultra-stable and adaptive systems*.

Finally, there was a Russian 'appreciation' of systems analysis as they believed it was invented and practised in the United States. This unique document presented some valid (and some invalid) assessments of where the professionals have been and where they may go. The article may be judged reasonable or unreasonable by the reader; but in any framework, it implies something about the state of systems analysis in the USSR.

1 Charles Hitch

An Appreciation of Systems Analysis

C. Hitch, 'An appreciation of systems analysis', The RAND Corporation, 1955, pp. 699, 8–18, 55, 1–25.[1]

Military systems analysis is an extension of operations research techniques of the Second World War to problems of broader context and longer range – e.g. force composition and development as well as operations decisions.

Greater complexity is inevitable as we attempt analyses to aid decisions affecting more distant time periods: growth in number of relevant variables; compounding of uncertainties; increased importance of enemy reactions; complications of time phasing; the need for a broader concept of criteria.

Techniques available to the systems analyst for dealing with these complexities are less than satisfactory. 'Factoring' is inevitable. All routine or mechanistic approaches are deficient; e.g. expected value and minimax criteria. Design and criteria problems are of dominant importance.

The systems analyst may have to be content with better rather than optimal solutions; or with devising and costing sensible methods of hedging; or merely with discovering critical sensitivities. He has an important role as inventor of systems.

The difficulties of systems analysis are rooted in the nature of the military problems. Other methods – e.g. piecemeal analysis; intuition – do not escape them and have limitations of their own. Military systems analysis provides a framework for combining the knowledge of experts in many fields to reach solutions which transcend any individual expert's judgment.

Components of systems analysis

The first widespread and explicit use of scientific method or analysis as an aid to military decision making was made by operations analysis (or operations research) teams in the Second

1. This paper is a condensation of lectures prepared for Air Force audiences; it was presented to the Operations Research Society of America at the Los Angeles meeting, 15 August 1955.

World War. Small-scale and partial use of similar analytic techniques had long been common. Thucydides describes examples of their use by the Athenians in the Peloponnesian wars.

But even Second World War operations analyses were limited in character. They related to operations in the immediate future, not to force composition or the development of equipment which would affect operations in the more distant future. Partly in consequence, they were simple, in the sense that they considered only a small number of interdependent factors. They were able, as a rule, to use some proximate, obvious straightforward objective or criterion as a basis for choosing one operation over another. A typical example of a Second World War operations analysis problem was what bomber formation to use in attacking targets deep in Germany. This problem had few variables and an obvious criterion: minimize losses in achieving given target destruction.

Since the war, at RAND and elsewhere, attempts have been made to use analysis as an aid to military decisions in problems of immensely greater difficulty and complexity. There has been a tendency to use the term 'systems analysis' to describe these more complex analyses, but there is no line of demarcation. Both operations analysis and systems analysis are attempts to apply scientific method to important problems of military decision, even though the problems are not particularly appropriate for scientific method and would never be selected for the application of scientific method by a truly 'academic' researcher. Both operations analysis and systems analysis have the same essential elements:

1. An *objective* or objectives which we desire to accomplish.

2. Alternative techniques or instrumentalities (or '*systems*') by which the objective may be accomplished.

3. The '*costs*' or resources required by each system.

4. A mathematical *model* or models; i.e. the mathematical or logical framework or set of equations showing the interdependence of the objectives, the techniques and instrumentalities, the environment and the resources.

5. A *criterion*, relating objectives and costs or resources, for choosing the *preferred* or *optimal* alternative.

The development of military systems analysis

Developments and extensions of operations analysis since the Second World War have taken the following forms:

1. The use of analysis to aid in force composition and development as well as operations decisions.

2. A great increase in number of interdependent factors considered.

3. The explicit treatment of problems of uncertainty.

4. The explicit treatment of enemy reactions.

5. The explicit treatment of time phasing.

6. A broader concept of objectives and criteria appropriate to the broader and longer-range problems of decisions being analysed.

We will discuss these in turn, considering in an impressionistic and superficial way the analytic techniques appropriate for each extension.

Force composition and development decisions

Models used for force composition and development decisions need not be formally different from those used for operations decisions. Their form is likely to differ, however, because of the greater number of interdependent variables, the increase in uncertainty, and the broader criteria appropriate to the longer time horizon considered. As an historical fact, the attempt to apply models to development problems has sparked important developments in systems analysis methods in all these areas.

More variables

The application to force composition and development decisions means that we are concerned with a military force two or three to ten or fifteen years in the future instead of with present forces. This fact alone vastly increases the number of interdependent variables which we have to consider. All sorts of things which are given in the short run become variables in the long run. For example, in the bomber formation operations analysis in the Second World War, the planes were B-17s, their number was given, the targets were given, the bombs were given, the enemy defences were given etc. In the longer run these are not given. They are unknown. They become variables. Some are variables subject to our control, some are variables subject to the enemy's control, some are subject to nobody's control. But all are variables. Moreover they are all interdependent.

In the Second World War bomber formation problem, on the same assumptions, everything else is given except possibly flight path. The number of cases, with the same degree of simplification, would be two, four, or at most sixteen. So just moving into the future puts us in a different ball park.

What do we do when we are confronted with a million or billion or decillion cases to compute and compare? Of course we develop higher speed computers every year and greater skill in using them, and I do not want to belittle the significance of this accomplishment. But even the capacities of modern high-speed computers are limited. We still have to cut down most of our problems for this reason. And there are usually more confining limitations. We have to get relevant data and relations to feed into the machine on all these systems. The more variables we deal with, the more time this takes and the more people – both scarce and valuable resources.

So somehow, for one practical reason or another, big, broad problems have to be cut down in size. We have to 'factor out' a practical problem – i.e. factor out those variables which are especially important for the decision with which we are concerned, and 'suppress' or 'aggregate' the rest. How do we do this? To a limited extent by preliminary analyses and tests, but for the most part by sheer judgment. It is hard to do; it amounts to no less than deciding, in designing the analysis, what is important and what isn't.

Probably most systems analyses that go wrong, go wrong here. Either they include a mass of data and calculations which are just excess baggage, or they exclude some really critical factor on which a good decision depends.

Explicit treatment of uncertainty

Uncertainty is present in operations problems, and it needs to be dealt with there. But uncertainties multiply, as an almost invariable rule, as we look further in future. We need to distinguish several kinds of uncertainty:

1. '*Planning factor*' uncertainty. Planning factors – attrition rates, average bombing errors etc. – are always uncertain, but less so in the present than in the future.

2. Uncertainty regarding the *enemy and his reactions*. As in 1, this is always with us and increases with time.

3. *Strategic* uncertainty. Will it be war, cold war, or peace? If war, when? General or local? With what political constraints? To achieve what political objectives? Who will be our enemies, who our allies? Will our allies make bases available to us?

In the very short run, strategic uncertainties of this kind are sometimes trivial – e.g. in the typical operations analysis problem in the Second World War. In the longer run they may quickly become dominant. No force composition or development analysis can ignore them. For example, the whole composition of the Air Force may be drastically affected by our opinion about the relative likelihood of big and little wars. Who can tell us? The answer is not easy or obvious. No present Administration or Joint Chiefs of Staff can make strategic decisions binding on a future Administration or Joint Chiefs of Staff.

4. *Technological* uncertainty. This is small in the present; in the future it can be vast. For example, until recently there was real uncertainty as to whether the H-bomb would work, and if so, when; this profoundly influenced the structure of many systems analyses. There is always technological uncertainty of some degree attached to research and development; otherwise it wouldn't be research and development.

5. Finally, there is *statistical* uncertainty – the kind that would still exist even if we could predict technological progress, the dates of wars, enemy strategies, and the central values of all important parameters. Statistical uncertainty stems from chance elements in the real world. Statistical uncertainty is the least of our worries in systems analysis: not only have we made great progress in Monte Carlo techniques for dealing with statistical fluctuations when we have to, but these are usually swamped by uncertainties regarding central values and states of the world in long-range problems. The use of elaborate methods to reflect statistical uncertainties in such problems is likely to be an expensive frill.

What do we do in a systems analysis to take account of the proliferation of uncertainties resulting from our ignorance of the future? We can't ignore them. To base our decision on some single set of 'best guesses' could be disastrous. For example, suppose that there is uncertainty about ten factors (e.g. will overseas bases be available? will the enemy have interceptors effective at 60,000 feet?), and we make a best guess on all ten. If the probability that each best guess is right is 60 per cent, the probability that all ten are right is one-half of 1 per cent. We would be ignoring futures with a 99·5 per cent probability of occurring. The problem breaks down into two parts:

1. How to compute all the 'interesting' contingencies. This raises problems very similar to those we have already discussed. Mere computation time is important, but usually much more so is deciding which cases deserve to be computed and assembling the data and structure for each.

2. Once we have computed the alternatives, we will almost inevitably discover that one strategy is superior in some contingencies, another in others. How does the systems analyst choose the preferred strategy? What decision does he recommend?

For example, suppose analysis shows that a strategic bombing system dependent on overseas bases will be most effective for 1960–65. Suppose further that we regard it as quite likely, but not certain, that we will have overseas bases in that time period. Suppose finally that if we should not have the bases, the system would be very bad. What do we do in such a case?

Ingenious efforts have been made to find an answer, but with limited success. The shortcomings of merely maximizing the *expected outcome* are too well known to elaborate upon: because it incorrectly weights high and low outcomes, it can lead to the choice of a reckless strategy and possible catastrophe. Maximizing the *expected value* or 'utility' of the possible outcomes is not subject to the same criticism of recklessness, but it is rarely possible to assign objective utilities to the outcomes, even if one somehow surmounts the theoretical and practical obstacles of assigning probabilities.

Expected outcomes or values, in any event, are appropriate only in playing against nature, not in playing against an intelligent opponent; and in few military systems analyses is it legitimate to ignore completely the possibility of intelligent opposition. Game theory suggests that in such circumstances we should *max-min*; i.e. choose the system which minimizes the worst that can happen to us. This would be good enough if we were really playing a two-person constant-sum game, but we never are – except perhaps at the lowest operational level. And if our opponent is anything less than completely wise and resourceful (i.e. always) *max-min*ing is too conservative; it forfeits opportunities to exploit enemy mistakes.

Some advocates of *max-min* have argued that a special variety of it – *max-min*ing 'regret' or 'loss' rather than outcomes – is an appropriate criterion or approximately rational rule of thumb for the case when we play against nature and can in principle calculate probabilities but in actual practice have only vague estimates. It at least does not lead to foolishly conservative strategies, as ordinary *max-min*ing does in these circumstances. But, as we have seen, we never or rarely find systems problems in which the game element is completely absent.

There simply is no satisfactory general answer to the problem. Different people take different views of risks – in their own lives and as decision-makers for the nation. Some play boldly, some play for safety. So what does the poor systems analyst do? He frequently calculates some expected outcome or minimaxes or both, but in interpreting his results he is aware of their biases and inadequacies. But he doesn't stop there. There are other tricks of the systems analysis trade where uncertainties are grossly important:

1. He tries to *invent* a system which is as good or almost as good if overseas bases are available, but still pretty good if they aren't. We call a system which is best in *any* circumstances a 'sure thing' or 'dominant'. We can seldom find a truly dominant system, but sometimes we can come close. Such a system in the above example might be a bomber force which would normally use overseas bases, but if they were not available would have a substantial capability with air refueling from ZI^2 bases. It is arguable that the most valuable function of systems analysis is the stimulation of systems invention.

2. If he fails to find a dominant solution, he calculates the cost of providing insurance against the chances of catastrophe – e.g. by buying a mixed force with a substantial number of very long-range aircraft. Then the Air Force has to make a command decision, but at least it can do so knowing what the insurance costs.

3. If he is concerned with *development* decisions, he recommends the development of aircraft and missiles only some of which depend on overseas bases. The situation may be clearer when decisions about quantity procurement have to be made several

2. Zone of the Interior, i.e. the area within the national or continental geographic boundaries [Editor].

years hence. We can afford to develop more types of equipment than we procure in quantity, and given the relatively low cost of development, we ought to. Insurance is cheaper at this stage. The systems analyst must guard against the implicit 'either/or'. Mixed systems frequently are right for procurement, even more frequently right for development.

Explicit treatment of enemy reactions

In some problems, what the enemy does is obviously crucial in making the right decision, e.g. the need for ECM[3], or the best allocation of funds between offense and defense. But it can also be crucial in many less obvious instances. For example, the choice between offensive bombers and missiles is extremely sensitive to their vulnerability to enemy attack – in the air and especially on the ground. So there is great interest in developing models appropriate for problems of this kind. Two kinds of models are available: *Game theory* models and *games* (i.e. war games).

Game theory is a branch of mathematics which studies situations of conflict. Two-person constant-sum game theory models are useful for some simple military problems; but satisfactory theories for non-constant-sum and multiperson games have not been developed, and the whole theory is in its infancy. Moreover, the difficulty of calculating solutions to game theory problems severely restricts the number of variables which can be included. Consequently there is a strong tendency to substitute games for game theory, and games have been developed which, unlike traditional military war games, permit many plays, so that a number of possible strategies can be tested against enemy counter strategies. Games, like game theory, are not completely satisfactory, but for other reasons. Different players play differently – some probably too well. Sensitivity analysis is usually impossible. Results are therefore very difficult to interpret.

Explicit treatment of time phasing

In many military decisions the *sequence* of events is of critical importance. For example, should we go into production now on some particular missile defense or wait two years until a better one is developed? To handle such problems we need 'dynamic' models, i.e. models in which the variables bear dates. We have such models, but introducing time explicitly is neither easy nor painless:

3. Electronic Counter Measures. [Ed.]

1. It complicates the computation by multiplying the number of variables. If we put time in we have to take something else out.

2. It complicates the selection of a criterion. Solution A may be better for 1958, worse for 1960; Solution B, vice versa.

3. It raises in acute form the question of our ability to predict, e.g. will a much better missile really be ready only two years later?

Broadening of criteria

The selection of a criterion or criteria is frequently the central problem of the design of any systems analysis. We have, characteristically, numerous alternative ways or 'systems' for achieving our objectives (positive values). All involve costs of some kind (negative values). If we had some common measure for all the positive and negative values (as a business firm does, more or less), the answer to the criterion problem would be obvious. You would choose the system which yielded the greatest excess of positive values over negative ones.

Unfortunately, we can't do this. Objectives and costs are usually incommensurate. Objectives are likely to be such things as enemy targets destroyed, enemy planes shot down, probability that war will be deterred. Costs may be crews lost, aircraft lost, millions of budget dollars expended. I don't want to exaggerate. Frequently we can find common denominators, especially on the cost side. But also frequently, especially on the objective side, things just won't add.

We have seen that in the typical Second World War operations analysis problem fairly simple, obvious ways out of these difficulties could usually be found. So many things are fixed or given. Thus in the bomber formation example, we can choose the formation which maximizes target destruction for given aircraft losses, or minimizes aircraft losses for given target destruction. These criteria, although they sound different, are logically equivalent, and so give the same answer (for the same scale of attack).

Consider, at the other extreme, the mixed force composition and development problem of determining the optimum composition of the Air Force, 1956 to 1966. What are Air Force objectives? It does little good to say that the Air Force objective is to promote such national objectives as to win a war if there is one, or to deter war. What kind of war? When? At what cost in lives and dollars? The optimal Air Force for fighting a thermonuclear war is not optimal for fighting peripheral wars. Concentration on missiles may be just right for our 1966 capability, but may weaken us for 1960. While our criteria must be consistent with national objectives, they must be defined much more concretely, or 'operationally'. No one has written (or will write), a real long-range war plan to guide us.

Of course, the criterion problem in the typical systems analysis is not as hard as in this case, which is extreme. It is extreme because a decision on force composition involves optimizing for the Air Force as a whole. It is a decision at the highest Air Force level. Decisions at this level are almost certain to require criteria based on complex and uncertain political and strategic factors. But many Air Force decisions, fortunately, can be made at lower levels – even force composition and development decisions. We can 'factor out' a problem and the variables which are of first-order importance in its solution, ignoring the rest. And instead of optimizing for the Air Force as a whole, we 'sub-optimize' for some sub-element of it, choosing an appropriate lower level criterion. For a fuller discussion of this point, see Hitch (1953).

Distinguishing problems which we can successfully factor out is an art. Let me illustrate with an example from SAC[4] operations. The choice between ZI and overseas bases (or what combination of both) won't factor. Overseas bases have tremendous ramifications throughout the Air Force and above and outside it. They have strategic and political values and vulnerabilities at other and higher levels than SAC operations that we would be stupid to ignore in making a decision. We can't sub-optimize, unless we frankly recognize that the sub-optimization deals only with some of the factors important to the decision.

On the other hand, given that we want at least part of SAC based in the ZI (and not even dependent on overseas staging), we think we can successfully factor out the problem: should range be extended by bigger, longer-range bombers, or by air refueling, or by what compromise between the two? This choice seems to have no first-order repercussions on air defense, on any theater operations, on international politics etc. It can be sub-optimized by choosing as an appropriate lower-level criterion the most efficient (in terms of target destruction) use of the budget or resources made available for strategic air.

But at best criterion selection is hard – harder in force composi-

4. Strategic Air Command. [Ed.]

tion than in operations problems, hardest in development problems. We have wrestled with criterion selection at RAND for eight years – with, at best, moderate success. We have found negative rules – criteria to avoid – but few positive ones of general helpfulness. There is no substitute for good judgment, and no substitute for exercising it. Working out a systems analysis with a bad criterion is equivalent to answering the wrong question. It is very easy to choose a criterion for a force composition or development problem that will insure our having the optimal system for the wrong war at the wrong time (to be fair, it is easy to make the same mistake without a systems analysis).

In some cases, indeed frequently, there is no single 'right' answer to the criterion problem. The ultimate values of decision-makers differ as well as their image of the future. In these cases the systems analyst simply has to conclude: if you want an Air Force which will do A, choose System X; if B, choose System Y.

The difficulties are in the problems

One may well ask, in view of this long catalogue of difficulties, dangers and limitations, and the rather obvious possibilities of abuse they open up, whether military systems analysis is worth supporting and continuing.

The first thing to stress in answering this question is that almost all the difficulties we have discussed are inherent in the nature of the military problems that systems analysis is designed to help solve, and by no means peculiar to systems analysis, however we define it. This is obviously true of the 'many variables' difficulty. In most of the military problems with which we are concerned lots of things just *are* important. Systems analysis has trouble including all of them. But so does any other conceivable approach. Systems analysis is increasing its capability to handle larger problems every year. With present computers, for example, we can deal with far more intricate models than a genius can manage intuitively.

Uncertainties make life difficult for the systems analyst, but this is so because the problem of intelligent behavior under uncertainty is *really* hard. Systems analysis at least permits one to explore systematically the possibilities of dominance and the cost of insurance.

Enemy reactions are hard for a systems analysis to incorporate. Game theory is in its infancy. War-gaming in any form has short-comings. But why? Because the problem itself is so hard. The enemy probably doesn't know himself how he will act or react in 1960. How can we outguess him? Not by abandoning game theory and war gaming, which for all their limitations carry us further along this road than any devices yet conceived.

Time phasing is so hard that few systems analyses attempt it explicitly. Why? Because of two difficulties: the choice of criterion when payoffs and costs occur in different years, and the prediction month by month and year by year of changes in technology and other parameters. Both are intrinsic. You can get rid of them only by escaping from reality. If the sequence of developments, or their speed, is crucial to your problem, you need an analysis with dates attached to every variable to find the right answer.

But note what a dynamic systems analysis can do even in the case where we can't predict the speed of development with accuracy. It can tell us what the critical development speeds are – if $> X$, wait; if $< X$, go ahead – and thus enables us to focus the intuition of experts on a manageable technical problem. It can even, in some cases, yield a surprisingly unequivocal answer despite rather gross uncertainties about speed of development. We have encountered such cases at RAND, in which, while postponement *might* have resulted in startling improvement in some performance characteristics, these were shown by systems analysis to have trivial military worth.

And finally, there are all those troublesome problems of criteria. I can only assure you, or remind you, that they are equally troublesome in policy discussions in the Air Council, in the Joint Chiefs of Staff, in the National Security Council, and in Congress. You don't escape from them by escaping from systems analysis. You may be able to fuzz them up or conceal conflicts in a clever debate or essay (you can do the same in a clever systems analysis), but they remain for a clever opponent to reveal. The fundamental difficulty is that there does not exist a clear-cut, definitive, operationally meaningful statement of national objectives or of Air Force objectives – even for the present, let alone for 1960–65.

The positive side

Before we can say anything in general about the usefulness of systems analysis we must know what we are contrasting it with. If we define systems analysis broadly to include the various game techniques, etc. discussed in the preceding sections, what are the alternatives? Let's consider two. Concentrating on the first word 'systems', the alternative is unsystematic or piecemeal consideration of problems. Concentrating on the second word 'analysis', the alternative is, I suppose, intuition.

Systems versus nonsystems

This distinction has nothing to do, necessarily, with analysis. It is a question of breadth of context. In principle, one can attempt to intuit answers in a broad or narrow context or use analysis in a broad or narrow context.

It would be foolish to maintain that broad systems contexts are good, narrow contexts bad. It all depends on the problem. Is it factorable or not? How factorable is it? Systems contexts can be too broad, and when they are, they are wasteful. You pay a heavy price for a broad context. For anything you put in an analysis, something must come out. The broader the context, the less detail. If you are a scientist trying to develop materials to withstand the heat of rocket engines, your chances of success will be reduced to the extent that you devote time and energy to pondering the relative likelihood of big and little wars. As a matter of historical fact almost all scientific and technological progress has been achieved within very narrow contexts – by scientists wearing blinders. Let's continue a fruitful division of labor and not all become systems analysts. Some of us are a little concerned that a large proportion of our best design engineers in aircraft companies seem to be spending their time designing systems analyses instead of aircraft.

Nevertheless, there are cases where the systems approach – the systematic examination of broad alternatives – throws a flood of light on important problems. Our previous example is such a case. What methods of range extension should be used to enable US bombers to reach targets deep in Russia? The broad alternatives are: overseas operating bases with medium bombers; overseas bases for staging only; big, very long-range bombers; mother–daughter arrangements; air refueling etc. Because the alternatives are broad, they need to be examined in a broad context. When we do we discover that some systems have a superiority of two to one to five to one over other plausible systems with enthusiastic advocates. This is a tremendously important conclusion. It could not have been reached, or at least not demonstrated, without a comprehensive systems approach and systems costing.

Let me give you another example of a different sort that cropped up in a RAND defense study of several years ago. It was at that time operational doctrine for certain interceptors to carry armament that, according to Air Defense Command estimates, gave each plane a 50 per cent probability of killing an intercepted bomber. Well, 50 per cent looked mighty good to most experienced Air Force hands. It was lots better than anything achieved in the Second World War. What did we find when we examined this doctrine in a systems context?

1. As was not the case in the Second World War, we were really preparing for defense against one (or at most a very few) massive atomic strikes.

2. The total systems cost of procuring and operating the interceptors – to get them into position prepared to fire a rocket at an incoming bomber – was extremely high, so high that the most lavish expenditure on armament scarcely affected the total.

3. It was therefore obvious nonsense to economize on armament.

4. The single pass kill probability and the kill potential of the defense system could be increased by nearly 50 per cent simply by increasing the armament. The performance degradation of the interceptor resulting from increased weight was of the order of 5 per cent, which, at least in the period of interest, had a negligible impact on kill potential.

Now this again was a tremendously important result of looking at a problem in a systems context. A systems *analysis* wasn't really necessary. I am sure that some Air Force officers, using a broad systems context, thought their way through this one and reached the right conclusion without so much as using the back of an envelope. But many apparently didn't, and the doctrine wasn't changed until the systems analysis was produced and presented. Systems analysis forces both the systems analyst and his audience to think the problem through in a systems context.

Analysis versus intuition

Let us turn to the second part of my comparison: analysis versus intuition. The main point I want to make might be called the inevitability of analysis. What we call intuition is a species of logical analysis. It uses models, in our sense of simplified conceptual counterparts of reality. Not surprisingly, in military problems as in so many others, it is sometimes useful to buttress our feeble minds with some external assistance: a pencil and the back of an envelope; a few equations; a desk calculator; sophisticated statistical and mathematical theory; high-speed calculators.

How far you go with such aids depends on the problem. But very frequently they enable you to find a solution you couldn't otherwise find, or to demonstrate that your intuitive solution was wrong – or, what is sometimes as important, right. I am not selling intuition short. The unaided human mind is quite remarkably proficient at solving some kinds of problems. Let me remind you:

1. Some human beings play very good games of chess. No machine can yet give them a match.

2. Human beings at RAND with intuition, pins and a piece of string found the optimal route in the Traveling Salesman problem out of 10 possible routes.

3. On the evening of last 2 November, the Columbia Broadcasting System used two methods for predicting the election results from the very early returns:
(a) The intuition of assorted political experts;
(b) A complex multi-variable model calculated on a UNIVAC.

The UNIVAC was grotesquely wrong: the experts did not do too badly. The human mind has some great advantages over any machine – if we think of them as rivals or alternatives. It has, by comparison, a wonderfully capacious memory, which enables it to learn from experience. It has a marvelous facility for factoring out the important variables and suppressing or aggregating the rest. Closely related to this facility, it can build models highly appropriate to the particular problem it is considering. Big formal models computed on machines are much less flexible. These are the reasons human beings beat machines at chess or war games. But, on the positive side:

1. It is utterly wrong to look upon intuition and analysis or minds and machines as rivals or alternatives. Properly used, they complement each other. We have seen that every systems analysis is shot through with intuition and judgment. We have experimented at RAND with man–machine combinations which will play war games better than either men or machines.

2. While unaided intuition is sometimes strikingly successful, as it was last 2 November or in the Traveling Salesman problem,

it can also fall flat on its face. For example, in the election of November 1952, when the UNIVAC was dead right, and the intuition of the experts so wrong that they suppressed the UNIVAC answer.

In contrast to the Traveling Salesman problem, try your intuition on this: there are twenty-five persons in a group. What are the chances that at least two have the same birthday? Almost everyone without statistical training says – very small. In fact, they are better than even. If there are sixty in the group, the chances are 99·4 per cent!

3. One of the troubles with intuition is that you never know whether it is good or not without an analytic check. For example, our intuition was good enough to solve the Traveling Salesman problem, but we didn't know it until we solved it analytically in a linear programming formulation.

MATS[5] didn't know its assignment of aircraft to routes was within 5 per cent of optimal until we worked out a linear programming solution.

I've said there are good chess players. But we don't really know. Maybe even the best are as far from optimal strategies as expert opinion so frequently has been on military problems.

4. Finally, analytic and computing techniques enable us to do things we otherwise couldn't. They may be poor on the memory side, but they have some capabilities unaided human minds don't.

Look again at the UNIVAC fiascos on election eves, 1952 and 1954. Here we had an elaborate model and a high-speed computer. In 1952 it was able to take the first few precinct returns from eastern states and trace their consequences – on its built-in assumption that similar trends were running in all precincts throughout the nation. They were, and the answer was dead right. The experts couldn't carry out such a calculation in their heads, and were inclined, like most experts, to err on the 'conservative' side.

Or look at a different kind of example – of insights derived from theory. Take a brand new theory – one that I described as in its infancy and of very limited usefulness – viz., game theory.

In connection with RAND defense studies we have long been interested in the optimal deployment of limited defenses among targets, some of which are more valuable than others. Unfortu-

5. Military Air Transport Service. [Ed.]

nately, we have found no satisfactory general rule for deploying defenses, but game theory has given us valuable insights and hints about good and bad deployments we would not otherwise have had.

One striking example: suppose you have your defenses deployed as well as you can. Now you get more defenses. How do you deploy them? Well, my intuition told me (and so did most people's) that you deploy them mainly to protect additional targets – additional cities, harbors, airbases etc. – that you didn't previously have enough stuff to defend.

Game theory says no. You use additional defenses mainly to increase the defense of targets already defended. In fact over a wide range, the more you have the more you concentrate it.

Having been informed of this startling result, you think about it and begin to see the rationale. An increase in your defensive strength is equivalent to a decrease in the enemy's offensive strength. But as his strength decreases, he has to concentrate more and more on your most valuable targets to achieve anything worth while. These are the ones on which you therefore have to concentrate your defense. But intuition alone would not have told us this. At any rate, not unequivocally enough to lead one to act on it.

Conclusion

Does analysis help more in the narrow context problems, where it has commonly been applied by scientists, or in the broad context problems, which are the special province of systems analysts?

I don't know. On the basis of results, certainly one would have to say that the case for analysis in broad context problems is comparatively unproved. Let me suggest, however, one reason why, when we are dealing with broad problems with broad systems analyses, explicit analysis using explicit models can be especially important.

We trust a man's intuition in a field in which he is expert. But in these case we are dealing with a field so broad that no one can be called expert. A typical systems analysis depends critically on numerous technological factors in several fields of technology; on military operations and logistics factors on both our side and the enemy's; on broad economic, political and strategic factors; and on quite intricate relations among all these. No one is an expert in more than one or two of the sub-fields; no one is an expert in the field as a whole and the interrelations. So no one's unsupported intuitions in such a field can be trusted.

Systems analysis should be looked upon not as the antithesis of judgment but as a framework which permits the judgment of experts in numerous sub-fields to be combined – to yield results which transcend any individual judgment. This is its aim and opportunity.

But we still have the question: where is the 'expert' in the field as a whole with the judgment required to design a systems analysis and interpret its results? We know there are not any real experts. But we think we can demonstrate that the degree of expertness required to design a systems analysis is less than the degree of expertness required to intuit a good answer without a systems analysis.

Let me put it in another way. We tend to be worse, in an absolute sense, in applying analysis or scientific method to broad context problems; but unaided intuition in such problems is also much worse in an absolute sense. Let's not deprive ourselves of any useful tools, however short of perfection they may fall.

Reference

HITCH, C. (1953), 'Sub-optimization in operations problems', *Journal of the Operations Research Society of America*, vol. 1, no. 3.

2 Malcolm W. Hoag

An Introduction to Systems Analysis

M. W. Hoag, 'An introduction to systems analysis', The RAND
Corporation, 1956, pp. 1678, 4–18, 56, 1–21.

A talk on systems analysis ought to begin with a definition of this
term. Unfortunately no precise, commonly accepted definition
exists. For the moment let me say merely that by systems analysis
we mean a systematic examination of a problem of choice in
which each step of the analysis is made explicit wherever possible.
Consequently we contrast systems analysis with a manner of
reaching decisions that is largely intuitive, perhaps unsystematic,
and in which much of the implicit argument remains hidden in
the mind of the decision-maker or his adviser.

Systems analysis is an outgrowth of Second World War
operations research, although it typically deals with choices that
concern operations farther ahead in time, and takes a somewhat
broader look at problems of military choice. Analyses at RAND,
for example, typically consider what equipment to procure and
develop for the Air Force of the future. Consequently the analyses
that are relevant differ a good deal from the operations research
of the Second World War. Dealing with the future, we are less
constrained by the specific equipments and modes of operations
that we happen to have at the moment. We have greater flexibility
and range of choice.

This difference between a Second World War kind of operations
research and systems analysis is less a matter of substance than of
degree. Consider, for example, the purchase of a house. If we are
interested in buying an old house, an associated decision about
furniture will be involved. One aspect of that decision may be the
choice of a refrigerator, and we may find that a space only thirty
inches wide exists in the kitchen for a refrigerator, that the house
is wired only for 115 volt current, and that no gas lines are
available. Consequently our choice of refrigerators is very
constricted, and for that reason the problem of choice may be
fairly easy. On the other hand, if we are buying a new house, one
yet to be designed, our choice of a refrigerator is quite a different

problem. If we sit at the drawing board with an imaginative architect, the kind of house we have, including a refrigerator, is wide open. We are no longer constrained to think the refrigerator must be no more than thirty inches wide, and we can consider the alternative of a gas rather than an electric refrigerator, or even an electric refrigerator that will utilize 220 volt current rather than 115 volt if that alternative is relevant. Under these circumstances, our range of choice is far broader and the number of alternatives that are relevant is consequently far greater.

Whether we speak of choosing refrigerators or military systems, some common conceptual elements are involved. Whenever we choose rationally, we try to balance the objectives we wish to attain against the costs of their attainment. In doing so certain common questions are always involved. First, what are the relevant alternatives anyway? Second, what in principle is the test of preferredness that we ought to apply in choosing among alternatives? In other words what is our criterion for choice? And, third, how do we go about the actual process of weighing objectives against costs in the selection among alternatives? To use a word that we will discuss a little later, what is our 'model' of the situation? How do we go about applying it, and how do we interpret the results?

Let us turn first to the question of the alternatives that are relevant. There are two classic errors that can be made here. Each consists of going to an untenable extreme position. Consider first the one error, which is to look at an unduly restricted range of alternatives. We should be making that mistake if, in terms of our homely illustration, we were to design a new house as if the refrigerator for it had to be constrained by all the specific limitations of an already existing old-fashioned structure. Of course, there are certain advantages in narrowing our range of choice arbitrarily in this way. It probably makes the analysis far more convenient. There are far fewer alternatives to consider, and we are less bedeviled by the problem of choosing. But we may pay a very high price for such a convenience. When we impose our constraints arbitrarily, some of the excluded alternatives may be far better than the ones left to choose from. This is a good way to get a bad refrigerator.

It is an equally good way to get a bad military system. Consider this possibility as an illustration. It is very convenient for the aircraft designer, torn by doubt about the size and other relevant characteristics of the payload that his airplane will carry many years hence, simply to assume that the physical configuration of a bomb will then be pretty much what it is today. Similarly it may be very convenient for the designer of bombs to assume that the bomb bays of the future will be of the same capacity as those of current models. But look what can result from such shared convenience. The airplane designer may believe that his airplane must be big in order to be able to carry a bomb as big as the current ones. In turn the landing space, whether it be fixed or floating, may have to be big. We end up with a combination of big bomb, big airplane, and big landing space. Yet it is quite possible that a more efficient system on the whole could be devised by combining a bomb small in size if not in bang with a small airplane operating from existing small airfields or carriers. Maybe such a better system would not be possible, but the point we want to make here is simply that whether it is possible or not can only be ascertained by looking at it. For that matter, an analysis that proceeded in such a narrow way would not look either at alternative systems that employed bombs much bigger in size than current models. If we look at an unduly narrow range of alternatives to begin with, we may never even pose the question in the right way. And getting a neat answer to the wrong question may be worse than an incomplete answer to the right question.

We can avoid the error of looking at an unduly narrow range of alternatives by specifying that we shall look only at broad systems for a comparison, and that all of the alternatives that are relevant in a broad system will be considered. In terms of the above illustration, we can try to look at all possible combinations of bombs of the future with airplanes and bases of the future. The great difficulty which this procedure raises, of course, is that of workability. The more inclusive the system, the more alternatives that are relevant, and the greater the difficulties of comparison.

Too few people appreciate the manifold alternatives that are available when one looks at a relatively unconstrained situation. Consider the simplest sort of arithmetic illustration. In asking what bombing airplane we ought to design for the future, let us concentrate arbitrarily for the moment on just three obviously pertinent characteristics – range, speed and payload. More of any one of these is obviously a good thing in itself. We should like to fly farther, faster and with a bigger payload. But, of course, we achieve none of these without cost. Indeed, in relevant cases we probably can achieve more of any one of these characteristics of performance only at the cost of the other two, let alone at the cost of still other considerations. If we want to fly faster, we must accept some range penalty, and so on. But suppose we have just these three characteristics to consider, and suppose, again quite arbitrarily, that to each of them we assign only a 'high' or a 'low' value. We can have a fast airplane or one not so fast, and so on. Now how many alternative combinations of these three characteristics do we have? We can answer this question by using the diagram given below in Figure 1. If we take any one of these characteristics as a beginning point, it can assume either of its two values – high or low. Each of these values in turn can be matched up with either the high or the low value of the second characteristic, so that we have four distinct combinations. Each of these combinations in turn can then be matched with either the high or the low value of the third characteristic. What we end up with are (2) (2) (2) or 8 combinations, as shown in Figure 1.

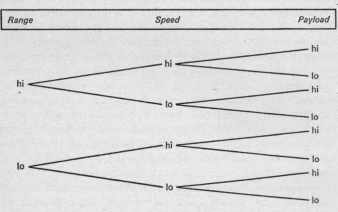

Figure 1 Combinations of values for three characteristics

The number of combinations of interest is not two multiplied by three, i.e. six, but rather two to the third power, which equals eight. The number of combinations increases exponentially, not by simple multiplication. Now that does not matter much when the difference is only between six and eight. But clearly we are interested in many more characteristics of our airplane. What about crew space, landing facilities and electronic gear, for instance? Yet if we include only ten characteristics of interest, and still restrict ourselves to but two values of each, the number of relevant combinations is (2) or about 1000. The difference between six and eight combinations is small, but the difference between twenty (two times ten) and 1000 is large. Typically, when we start with clean drawing boards and are free to consider all alternatives of interest in a problem of the future, there are many millions of possible alternatives. The problem is one of choosing sensibly from among that large number. Given that kind of a situation, it is manifestly absurd for anyone to say that we should develop all alternatives in a hardware form, see how they perform, and then choose among them. It is imperative to find a cheaper way to compare them. The bulk of the alternatives must be excluded by a comparison that utilizes few, if any, tools beyond pencils, paper, and discriminating thought. Given that this is the situation, there is surely a case for making the choice on paper as carefully and as ably as we can. Hence the desirability of systems analysis.

You notice that all I have done in discussing the range of alternatives is to point up the error of going to either extreme. The unduly narrow analysis may exclude the really interesting alternatives. The impossibly broad comparison is simply out of the question. Nobody but a fool or a charlatan ever pretends to look at all the universe and to resolve all problems at the same time, even though in principle those problems are interdependent. In a general discussion there is very little that can be said positively about how big the context of a problem ought to be. It depends on the particular problem at issue. All one can say is that the analyst must be very careful to make his analysis neither too small nor too large, which is only to say that judgment is essential at this very early stage.

The problem of the criterion, or test of preferredness, that ought to be used in a particular problem is always with us. To take one aspect of our military problem as an example, let us consider the question of the preferred engine for the bomber of the future. Suppose, again arbitrarily, that only turbo-jet engines are at issue. One possible criterion for choosing among alternative engines is that of pounds of thrust per pound of engine weight. Clearly, greater thrust is desirable; equally clearly, light engines are desirable. We can balance the one characteristic against the other for the alternative engines that we think we could have. The result of a comparison might appear as in Figure 2 (on page 42).

On the basis of this comparison alone we should prefer engine A to engine B. But any sensible engineer will immediately

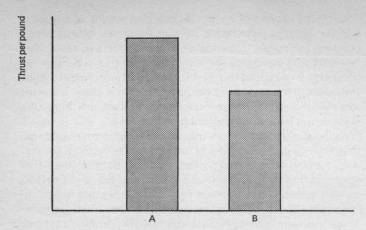

Figure 2 Comparisons of thrust per pound and engine type

object that this is too simple a criterion for engine choice. It leaves out of account many important characteristics, and there is no assurance that these other characteristics will influence the decision in the same direction as the two characteristics of thrust and engine weight that are considered. An engineer will immediately ask, 'Is engine A more durable than B? Is its specific fuel consumption lower? Is it cheaper to produce?' Certainly these are relevant questions, and they do not answer themselves. It is possible that engine B is decidedly superior to A with respect to one or more of these additional characteristics, and there are undoubtedly other relevant characteristics as well.

Moreover, we may note in passing that there is one other question that certainly ought to be asked. What size of engine in absolute terms results from following such a criterion to the bitter end? It is possible that we would get a 60,000 pound thrust engine, one so big that we might not be interested in it at all. The use of a ratio as a criterion for choice yields no assurance that the absolute scale of the operation in question will be an interesting or relevant one.

Given the difficulties that we can get into with an obviously incomplete criterion, what are we going to do about it? The only way to resolve the question is to look at the comparison in terms of the context of a broader system where we can bring a more complete criterion to bear upon the problem. For example, we can consider the context of a strategic bombing system as a whole. Suppose we take a 'reasonable' list of Soviet targets for a strategic bombing campaign, and stipulate that the objective of all the systems to be compared is to acquire as great a capability of destroying those targets as possible for a given total budget. (You will notice that I avoid all question of what consitutes a 'reasonable' target.) What system that can be devised will have the greatest capability given the budget constraint?

As we experiment in our paper comparison with different bombing systems, we are free to choose among engine types. We shall try to choose that engine which, when married to all the other desirable components of the bombing system, is consistent with the greatest combat capability. The issue between engine A and engine B will now tend to be resolved. If we only listed the technical characteristics of engines A and B, it is likely that we would have two columns: one column of characteristics in which engine A was the superior of engine B, and a second column of different characteristics in which engine B was superior to engine A. So expressed, there is no common denominator in terms of which of these different characteristics can be compared. Such a listing would suffice for making a choice only if one engine uniformly dominated the other, that is, was better in each characteristic. One just does not expect to find that sort of dominance in an interesting problem. Interesting problems are typically not that easy. Therefore, in order to choose between the two engines, we must find a way to reduce their diverse merits and demerits to a common basis. The context of the strategic bombing system as a whole makes it possible for us to find that common denominator. Each merit of a particular engine type tends to have its impact in terms of a contribution toward greater target capability. Conversely each demerit tends to have its effect in terms of decreased capability. The net effect of the merits and demerits of each will result in a final contribution to total system capability. Their differences can be expressed in terms of the common denominator of targets, and which engine is better can be established given a budget constraint.

How can we determine the appropriate scale of the bombing campaign at issue? How big a budget should we impose as a constraint? It may well be that we cannot answer this question with any precision. In that case all that we can do is to try several budget levels and see what happens at each level. If we are lucky we find that our preference for one engine type over the other does not depend upon the budget level. That is, choice between the two is not sensitive to variation in the scale of the campaign. If so, we can make a strong recommendation for that engine.

The criterion in our bombing campaign comparison is expressed in terms of maximum target capability for a given total budget. The criterion could equally well be expressed the other way around in terms of minimum total budget for a given target capability. At any particular scale these two expressions normally amount to the same thing. If we express the criterion the one way, we get a result like that shown in Part 1 of Figure 3; if the other way, we get a result like Part 2 of Figure 3. In either case the margin of superiority of systems employing engine B over systems employing engine A is much the same. The result of our comparison in the context of a broader system is to correct the misleading impression of a technical superiority of engine A over B that was generated by the overly simple criterion of maximum thrust per pound of engine weight (Figure 2).

The comparison in Figure 3 is in terms of a scale of a $1 billion budget (Part 1) or 100 targets (Part 2). As we try different scales of effort, notice what happens. At each scale that we try, say $1 billion, $2 billion, and so on, we shall try to establish that system with maximum target capability. Once that system has been found, its cost can be transposed to a different chart that will graph the relationship between the number of targets that can be destroyed and the total cost necessary for their destruction. We end up with the chart in Part 2 of Figure 4.

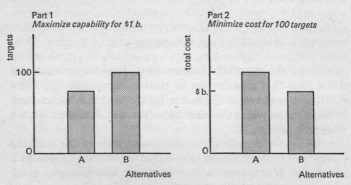

Figure 3 Alternative criterion wordings
Part 1 : maximize capability for $1 b.
Part 2 : minimize cost for 100 targets

Incidentally, such a chart serves to demonstrate the fallacy of one commonly encountered criterion. What does it mean if somebody says the criterion for this job is to get the most for the least? The 'least' in this case means lower dollar cost, which means we move to the left in Part 2 of Figure 4. But getting the 'most' means that we destroy the most targets, which means that we go up. But we cannot go up and to the left simultaneously in this chart. Higher capability means higher cost once we have determined the efficient systems to be employed at various scales of effort. Such a criterion involves us in a simple verbal contradiction. To convert it into a meaningful proposition, we must rephrase it by saying, 'Get the most for a stipulated cost; or, alternatively, for a stipulated objective get the lowest cost that you can.'

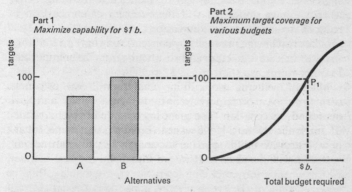

Figure 4 From one scale to many
Part 1 : maximize capability for $1 b.
Part 2 : maximum target coverage for various budgets

So far I have ignored the vital matters of how you go about the actual process of comparison in a systems analysis. We must simulate a real life performance by alternative systems, and naturally we want to make the simulation as realistic as possible for the important factors at issue. I stress the word 'important'. It is very clear in any analysis that we cannot take everything into account. Something must be left out in order to make the analysis workable. Given that we have to eliminate something, we shall try to eliminate the less important.

Sometimes the problems of our military analyses are compared to the problem of the owner of a racing stable who wants to win a horse race to be run many years hence, on a track not yet built, between horses not yet born. Everybody would agree that a prediction of the kind of grass that would grow in the infield of the race track is quite unimportant in the analysis, and need not be included. But somebody would certainly quarrel with a failure to consider the composition of the running surface itself. Race track fanciers tell me that some race tracks are faster than others, and that some horses perform relatively much better on slow tracks than they do on fast tracks. Again we only point to something at once important and difficult; we do not tell you how to do it. How do you decide which factors to suppress and which ones to include? And yet an analysis not only of necessity suppresses some things; it aggregates many of the obviously important things that remain. A multiplicity of small things must be aggregated into a small number of manageable big things. Only when we have done this will the problem of comparison be reduced to manageable dimensions. Consequently we shall simulate real life incompletely by concentrating only on the things that we think are important and abstracting from the unimportant. In other words, we shall create a 'model'.

Most of the actual work in any analysis will deal, of course, with the choice of factors to include in the model and their quantitative estimation. The great importance of careful, skilled estimation is obvious. If we want to build the best home, we ask how alternative building materials compare in strength, weight, aesthetic appeal and cost. If we want to build the best bomber, we ask what performance characteristics are attainable within the expected state of the art in aircraft construction, what relevant trade-offs can be made among those characteristics, and what defenses future bombers may have to face. There are many more relevant questions that could be listed in either case. All must be answered as best they can be in order to make a comparison possible, and the better the answers, the better the analysis as a whole. There is little more to say at a level of general principle except that it is important that the various degrees of confidence with which particular factors are estimated should be expressed, and the variance attached to particular estimates is important as well as the best guesses about the expected magnitudes.

Given a common denominator with which to compare physically diverse things, all or a good part of the model will typically be quantitative in character. Hence the talents of the mathematician, of value elsewhere in the analysis, are indispensable at this stage. We turn to a mathematician not so much for his skill in performing mass computations as for his ingenuity in escaping them. With judgment, one may be able to boil down an unworkably large number of alternatives to a number which, while formidably large, is manageable. With ingenuity, one may be able to substitute simplified for intricate mathematical manipulation. When all of the honest tricks of the trade that permit such reduction have been utilized, we have the electronic calculator to fall back upon to handle masses of computations. Improvements in the art of mathematical manipulation and the development of the electronic calculator have contributed greatly to the potential for useful systems analyses.

So far we have talked about several common conceptual elements that enter any problem of choice. But behind all of them is one troublesome element of which you are all aware. When we talk of the future we are speaking of a future largely unknown. Any analysis of current decisions that will bind us in the future is plagued by the difficulties raised by uncertainty. For military studies, what will the enemy be like in the future? How good are our forecasts of the technological improvements that will be possible if we will only give the go-ahead signal now to certain development and production programs? And in the event our military devices are actually employed in the future, what chance elements will help to determine their success? How lucky or unlucky will we be in the particular situations?

We are uncertain about all of these things, and our uncertainty gives us a two-fold problem. Because we are so uncertain about the magnitudes of many of the important factors, we don't know how to push a comparison between alternatives to a definitive conclusion even when we have a definite criterion for choice between them. As if that were not bad enough, the formulation of a criterion for choice in uncertain situations may itself be elusive. To take the simplest sort of example, different people may give different answers to the question of whether they would prefer a certainty of $5 to one chance in twenty of getting $100. Yet the mathematical expectation of gain is the same in both cases. We cannot fix one definite criterion for choice between these two that will apply to all individuals.

Granted that uncertainty tends to be pervasive, what can we do about it in systems analysis? Well, first we can do the same sort of thing we did when faced with uncertainty about the budget level in our example. Once we have a particular result based upon one set of assumptions, that result can be tested for its sensitivity to changes in many of the factors that enter into the analysis. We can try an altered scale of effort by assuming a different budget, we can double the strength of the estimated enemy defenses, and so on, and see what such tests do to our choices. One essential of most systems analyses is a battery of sensitivity tests that show how the results depend upon different factors. If we are lucky, we emerge with a result that is comparatively insensitive to a wide range of change in many critical elements of the analysis. If we are unlucky, we emerge with at best equivocal recommendations, although the factors upon which choice critically depends, and the particular uncertainties that it is important to try to resolve, will have been clarified.

For dealing with uncertainty, some kinds of statistical uncertainty can be treated explicitly in the model. This possibility raises technical matters that I do not propose to go into here beyond noting that the criterion can be expressed in a somewhat different form. Instead of seeking the minimum cost system among systems all of which have an expectation of attaining a certain capability, the criterion can be rephrased, for example, in terms of a confidence level: 'Choose the least costly of the systems that have a 90 per cent assurance of being able to do a stipulated job.' Given this different sort of criterion, different model manipulations may be called for, and one may get a somewhat different solution.

Finally, and probably most important of all, we can try to build into our analyses a provision for hedging our bets. We can try to protect some of the bets we make upon an uncertain future by making others in opposed directions. In that way we may not be caught out too badly by particular shifts of fortune. For example, if we are making recommendations for novel types of equipment at quite an early stage of development, we may want to recommend multiple programs: not 'A is preferred to B', but 'develop at least both A and B from among a much larger number of possible alternatives.' If both are developed, we are much less exposed to catastrophe than if we had plumped all out for only A or B. The chance that both will be failures is perhaps much lower than that either will be a failure if we concentrate exclusively upon it. To revert to our future horse-race analogy, one may want to analyse carefully in choosing preferred breeding stock, but prefer to seek more than one colt from more than one set of parents. But not, of course, without limit. Stud fees and the prices of brood mares are high for blood lines of proved merit. And modern hardware development for military application is not cheap either.

The principle of hedging can be extended all the way through to force composition and employment decisions. One may argue cogently for systems of mixed weapon types on the basis that we are less uncertain about getting tolerable results than in the case of a pure system of one type, even in cases where our best guess of probable results tends to push us toward a pure system. But the cost of this type of hedging rises as we move from the early stages of research and development toward procurement and use, and we can afford less of it. Moreover, as we move from early to late stages of development the burden of uncertainty lessens, and hedging loses some of its great value. Incidentally, this consideration suggests that we supplement our hedges by not making particular decisions prematurely. If there is no necessity to choose an engine now because the weapon program at issue is bound to be held up for a long period of time anyway by the need to develop a guidance system, it may be very desirable to defer our engine choice to a time when our decision can be a better one.

You notice that none of these comments tells you precisely what to do in particular situations. It is very clear that systems analysis as currently practiced, and probably as practiced in the future, is much more an art than a science. To be sure, the analyst ought to employ whatever available scientific tools are appropriate at particular places in the analysis, but his operations as a whole are not characteristic of 'Science' with a capital 'S'. We have stressed the part that judgment plays at each stage of a systems analysis. What scope should the analysis take anyway; what is the appropriate balance that escapes being over-ambitiously big or unduly constricted? What criterion ought to be applied? What factors can be suppressed in the analysis? Of those that remain, what degree of aggregation of particular factors is permissible? How are they to be manipulated? How is the result to be interpreted?

Moreover, our very division of systems analysis into 'stages' has too much of a cut-and-dried flavor, although that division is useful in order to talk about some conceptual issues one at a time. In practice nobody is likely to give you a problem neatly and correctly formulated, with the identification of relevant alternatives a matter for simple enumeration and the other stages to follow. One of our senior practitioners of the art of systems analysis insists that the term itself is unfortunate and should be replaced by 'systems design'. In his view the perception of the problem and the finding of relevant alternatives are the key creative parts of a study, and the use of the term 'design' is meant to highlight those parts. The imagination to perceive an important problem before others have become aware of its imminence, and the inventiveness to devise novel alternatives for its solution, are perhaps the most important qualities the systems analyst needs. And, like the sound judgment also required, these qualities cannot be taught in terms of general principles or set rules. One can only repeat admonitions: 'systems analysis is no substitute for sense', and 'There are neither prescriptions nor substitutes for ingenuity in analysis'.

Sometimes when stress is properly put upon all the difficulties that preclude a definitive analysis of decision problems in terms of a full-fledged scientific discipline, one encounters the following sort of reaction to a study: 'There are so many imponderables in the situation, so many uncertainties, that I just have no faith in the results that emerge from all this mass of data and discussion. I prefer to avoid it, and therefore I avoid systems analysis.' There is a direct answer to this type of understandable objection. The difficulties are intrinsic in the nature of the problems that are studied. They are *not* difficulties that are grafted onto the problems just because particular techniques are brought to bear on them. The same burden of uncertainty must be borne by anyone who tries rationally to arrive at decisions. The man who solves complex problems in the span of five minutes on the intuitive basis of 'sound' military or business judgment and experience, and that is certainly possible, may then forget about it and sleep easily at night. In contrast, the man who has worked two years on an involved systems analysis of the same problem may sleep badly, but only because he has become acutely aware of all the pitfalls in the problem.

At RAND we tend to have faith that systems analysis can do a much better job on many problems than conventional staff work in the military organization, or, for that matter, elsewhere in government or industry. If you take bright people who are scientifically trained, who are detached from the particular problem and consequently have no bias, and give them ample time and opportunity to bring their combined diverse experience to bear upon the problems, they certainly ought to be able to do considerably better than one harassed Indian who is given, say, three months to solve an impossibly big problem with little assistance. And they ought to do better even when the comparison is a fairer one.

One should not overstress the difficulties of finding the best solution to complex problems by analysis, even though the difficulties are great. To be sure, many problems by their very nature are so complex that one would never be confident that the best result had been found. At the same time very useful results may be achieved. In particular, one may tackle a problem in order to find a better rather than the best solution. Instead of starting with a clean drawing board, looking at all the relevant alternatives that can be discovered, and endeavoring to find the best, one may take as a starting point the existing program or pattern of decisions that, unless shortly altered, will bind the future. Taking that as a starting point, in what respects does the program seem to be demonstrably bad? If such respects can be found, one can proceed to the demonstration that they are bad. One seeks an alternative program that can be expected to do better in the likely contingency, and yet be better adapted to meet unlikely but possible contingencies. Probably we would be better advised in many problems to tackle the design of better systems rather than the abstract analysis of the best. If we did, we might be able to make useful recommendations that would otherwise escape us, for we should bog down in inextricable difficulties if we pursued the best of all possible worlds. At the worst, definitive recommendations may not be possible, but analysis will at least make more explicit for the decision-maker many elements pertinent to his decision of which he is imperfectly aware. At the best, one may do far better. Demonstrably important and secure recommendations may be produced.

It is only fair to add that while systems analysis can and ought to be much better than conventional staff work, it can be worse. Many systems analyses share an impressive facade. The technical discussion is long and complicated, the charts are elegant, the mathematical appendices are formidable, and there is great display of technical jargon and virtuosity. Such a facade can reveal very good analyses or it can conceal very bad ones. If it conceals bad ones, it may be worse than no analysis at all. The difficulties of understanding the technical display may impede rather than facilitate relevant criticism. The whole elaborate structure can conceal a bias in terms of arbitrarily excluded alternatives, an improper criterion, bad execution of the analysis, or poor interpretation of the results. But all tools can be used or abused. The possibility of misuse need not and should not deny us the opportunity to make good use of the available tools. After all, 'The substitute for bad analysis is good analysis.' How else do you know it is bad?

3 N. Jordan

Some Thinking About 'System'

Excerpts from N. Jordan, 'Some thinking about "system"', The RAND Corporation, 1960, pp. P. 2166: 1–31.

Prefaces alternative definitions of a 'system'
From Webster's New International Dictionary, unabridged second edition

1. An aggregation or assemblage of objects united by some form of regular interaction or interdependence; a group of diverse units so combined by nature or art as to form an integrated whole, function, operate, or move in unison and, often, in obedience to some form of control; an organic or organized whole; as, to view the universe as a *system*; the solar *system*, a new telegraph *system*.

Knotted *systems* of steep, small hills. *Owen Wister*.

2. **Specif.** (a) The universe; the entire known world; – often qualified by *this*; as, to regard *this system* with wonder. (b) The body considered as a functional unit; as, malaria pervades *this system*. (c) Colloq. One's whole affective being, body, mind, or spirit; as, his insinuation finally got into my *system*.

3. An organized or methodically arranged set of ideas; a complete exhibition of essential principles or facts arranged in rational dependence or connection; as, to reduce the dogmas to a *system*; also, a complex of ideas, principles, doctrines, laws etc. forming a coherent whole and recognized as the intellectual content of a particular philosophy, religion, form of government, or the like; as, the theological *system* of Augustine, the American *System* of government; hence, a particular philosophy, religion, etc.

'Our little *systems* have their day.' *Tennyson*.

4. Hence: (a) A hypothesis; a formulated theory. (b) Theory as opposed to practice. (c) A systematic exposition of a subject; a treatise. *All now rare*.

5. A formal scheme or method governing organization, arrangement etc., of objects or material, or a mode of procedure; a definite or set plan of ordering, operating, or proceeding; a method of classification, codification etc.; as, the Dewey decimal *system* of classifying books; the Berillon *system* of fingerprinting; the Belgium *system* of tunneling;

according to the Linnaean *system*; seeking a *system* by which to win at roulette.

6. Regular method or order; formal arrangement; orderliness; as, to have *system* in one's business.

7. (Usually with *the*) the combination of a political machine with big financial or industrial interests for the purpose of corruptly influencing a government. *US*

8. **Biol.** Those organs collectively which especially contribute toward the important and complex vital functions; as, the alimentary or nervous *system*.

9. **Engl. Law.** Method or design as shown by other acts of a dependent similar to that charged, evidence of which is admissible to rebut or negative a defense of accident, mistake, or ignorance, or to prove a course of conduct.

10. **Geol.** A division of rocks usually larger than a series and smaller than a group, and deposited during a period; as, the Silurian *system*.

11. **Gr. and Lat. Pros.** A group of two or more periods. Also a group of verses in the same measure.

12. **Music.** (1) An interval regarded as a compound of two lesser ones; – so used in Byzantine music. (2) A classified series of tones, as a mode or scale. (3) The collection of staffs which form a full *score* (which see).

13. **Physical Chem.** An assemblage of substances in, or tending toward, equilibrium. *Systems* are classed as two component, or binary; three component, or ternary; etc.; also as univariant, bivariant etc. (See *Phase Rule*).

14. **Transportation.** A large group of lines, usually of somewhat diverse character, under common ownership or permanent common control; as, the New York Central *System*.

15. **Zool.** In many compound ascidious, a group of zooids arranged about a cloacal cavity serving for them in common and into which the atrial orifices of all open.

Introduction

Recently I experienced something that struck me as quite strange. It was my good fortune to be able to attend an interdisciplinary meeting devoted to the study of and discussion on 'self-organizing systems'. The participants at the meeting represented a gamut of disciplines ranging from the physical and engineering sciences through the physiological and medical sciences to the psychological and sociological sciences. Eighteen people addressed the meeting and many others participated in subsequent discussions. Every one of the people who spoke found it difficult, either explicitly or implicitly, to define 'system' in general and 'self-organizing system' in particular. In fact, one very senior and influential scientist in this area asserted that there is no such thing as a self-organizing system; then he asserted that he would continue to use the term – and did so.

Nevertheless, the meetings were quite successful as these things go. The papers were well thought out and, by and large, challenging. The animated discussion that followed them showed that they generated serious thinking on the part of many of the audience. This is paradoxical. For how can we speak intelligently, interestingly about something of which we cannot think clearly, which we cannot define?

Since then I have heard many people speak about problems concerning systems at various occasions. Forewarned, I was on the alert. It is a fact that many who speak about systems are uneasy with 'system'. They assert that they will not try to define it, that it is vague, ambiguous, fuzzy and even meaningless. And yet, since both the speakers and the audience do have a concrete system in mind, the subsequent discourse using this indefinable term proves often to be valuable and rewarding. 'System' is, of course, now defined as the concrete system under discussion. This permits the ensuing discussion to be fruitful, but it also has some undesirable side effects.

We cannot use words cavalierly, as Humpty-Dumpty recommends, without reaping some crop of confusion, be it in communication with others, or in communication with ourselves. All of us are aware of most, if not all, the definitions of 'system' given in the dictionary – the definitions which are reproduced in the preface to this paper. What's more, intuitively, willy-nilly, we accept all of them; after all, we have no alternative. The word is actually used in these ways in intelligent, meaningful discourse. Asserting that what we mean by 'system' is just a concrete concatenation of elements, the topic of the discussion, does not help us in the ensuing dilemma which makes every other definition of 'system' either wrong or meaningless. We cannot use words arbitrarily without, concomitantly, sapping the foundations of the organized, meaningful, stable world in which we live; in which we must necessarily perceive ourselves to live in order to function effectively. The apology all too often heard that 'system' cannot be defined is a symptom of the disquiet, the ill-at-easeness that such a sapping generates.

Man cannot function too effectively when he is ill at ease;

neither can he think too clearly. Discomfort generated by semantic confusion is generally unnecessary. This paper will attempt to dispel at least some of this confusion involving 'system'. It will attempt to show that 'system' is semantically legitimate *per se*, despite its many specific meanings, by:

1. Reviewing some obvious but neglected facts of perception and cognition.

2. Explicating an implication of these facts – the 'core meaning' of a word.

3. Trying to show that the many definitions of 'system' are correct applications of the core meaning of 'system' to concrete cases.

4. Formulating a possible taxonomy of these applications.

Some obvious but neglected facts of perception and cognition

To say that something is obvious often serves as a kiss of death so far as further consideration of that something is concerned. But the obvious is often anything but simple. And in attempting to unravel the complexities that underlie the obvious, insights may be reached which make something else that appeared to be complex and subtle thereby simple and obvious. The obvious upon which I wish to focus attention is the nature of a thing, the nature of a stimulus, the nature of an idea. In what follows I will restrict myself to vision, but the points to be made are relevant to perception in general and are independent of the specific sensory modality mediating the perception.

In vision the lens in the eye focuses a projected picture of the environment upon the retina. The retina itself is basically a two-dimensional plane of points, each point being a light-sensitive terminus of an individual neuron. Hence, in the visual perceptual process, there is a stage where the non-homogeneous pattern of light rays which 'carry' the picture of the environment is transformed to a non-homogeneous stimulation of a set of points that constitute the retina. As a result of this stimulation, each neuron that has a terminal point in the retina undergoes an electro-chemical change which starts a process that goes 'upward' into the central nervous system. As an end result of the process, the person sees the environment.

Somehow, and at the present mysteriously, the organism manages to partition this set of discrete point stimulations into two sets, one set becoming the visual figure being looked at and attended to, the other set becoming the background to the figure. The figure looked at is experienced as being 'a thing'; whereas, the background is just sort of there, and may consist of 'nothingness' or some combination of 'nothingness' and things unattended to.

The problems entailed in how the nervous system functions are well known to psychologists and physiologists, and much work is current in a quest for solutions. There is, in addition, a closely related problem which is not attended to – the ability of man to see many different things in the environment confronting him. In other words, man can and does partition the set of discrete point stimulations relatively freely. This enables him to see that the figure has discriminable parts or that the background has discriminable parts. Even more important, this ability enables him to shift from one figure to another without a change in the visual field. Man, therefore, has the ability to organize the punctiform neural stimulation pattern and reorganize it. And this is important; both the organization and the reorganization are to a great extent a function of what is of interest to a person at the time of perception, of what is relevant to him.

The world appears as it does because we have a sensorium that responds to certain physical stimuli and a central nervous system that can process these responses and organize them in determinate ways. It so happens, and not by chance, that these ways are biologically relevant and this facilitates man's existence on the earth. The above analysis may point to a solution of a vexing metaphysical problem which plagues and has plagued many philosophers as to what things really are. The 'fundamental' things which populate the world we live in are those aspects of the world which can generate punctiform stimulation patterns upon our sensorium which are organizable into visual figures. Since the same visual field can be organized and reorganized in various ways depending upon vagaries within the perceiver, it seems proper to speak about *modes of visual organization*.

Something similar seems to hold for cognition; by cognition I mean what is generally denoted by 'thinking'. The information stored in the 'memory banks' of a person's brain is analogous to the punctiform stimulus pattern impinging upon the peripheral sense organs. The idea we think about or the thought we are considering is analogous to the visual figure.

In olden days people liked to think that for each idea there corresponds an engram of a physical nature which is plucked out into consciousness when needed. Even slight consideration in a restricted mode of thinking, the use of words, shows that this is not feasible. Consider a person speaking or writing. Meaningful sentences emerge full-blown. These sentences are combinations of words. Many of these sentences are unique in that they have never been used before. A sentence, by definition, expresses an idea. Yet the sentence *per se* could not have existed as an engram as such before its formulation. Words, of course, do and did exist; in this example, words are analogous to the punctiform stimuli. The sentence which expresses the idea is an organization of words. And this organization is determined by what is of interest to the person at the time of expression, of what is relevant to him.

It is difficult for me to grasp how an idea expressible by a sentence can exist as a determinate engram while the sentence expressing it must be organized from the unit words in the person's 'memory banks' at the time of expression. It seems simpler, and probably more correct, to look upon the process of organizing the words as one aspect of the basic underlying process of organizing the idea to be expressed. It is possible to bring additional arguments to buttress the contention that cognition exhibits rules of figure-ground organization and shifting figure-ground relationships similar to those found in perception, but I feel the above example to be sufficient.[1] Hence, consistent with the perceptual analogue, it seems proper to speak about *modes of cognitive organization*.

Perceptual figure-ground organization and its shifting relationships seemed to point to a clarification of the metaphysical problem of what a thing is; conceptual figure-ground organization and its shifting relationships seems to point to a clarification as to what creativity is. Given a set of punctiform elements of some sort dwelling somewhere in the brain it is relatively easy to conceive of creativity as being a novel organization of a subset of these elements which, in turn, may lead to the creation of a new element. It is much more difficult to conceive of creativity as the emergence of a new engram from, from . . . from where?

1. 'Figure-ground organization' and 'shifting figure-ground relationships' are technical names for the two perceptual phenomena discussed above: (1) the ability of the organism to segregate the punctiform stimulus manifold on the retina into a visual figure and a background; and (2) the organism's ability to segregate the manifold into various different figures and backgrounds depending upon, among other things, its interests.

On words and substantives in general and on 'system' in particular

It is the vogue among some philosophers and scientists to assert that the spoken language is a poor conveyer of meaning, that words have fuzzy meanings, ambiguous meanings and sometimes even self-contradictory meanings. For evidence in support of this assertion they can point to many instances where people use words in a fuzzy, ambiguous or self-contradictory way. At this the intelligent man in the street will indignantly rise and 'frumiously' exclaim that in these instances people misuse words, and that words should not be blamed for their misuse. And he is right. One should not blame words for the unfortunate fact that people cannot use them correctly.

But, as is often the case, the philosophers and scientists are right too. Common language is a poor conveyer of what they have postulated meaning ought to signify. They have postulated that the world must consist of some given set of irreducible units and that all perceivable and conceivable phenomena must be a result of lawful combinations of these units. A rigorous language should then be of a nature such that a sentence constructed in that language which denotes a phenomenon should clearly indicate the units and the lawful combinations which go into generating that phenomenon.

It so happens that regardless of whether these postulates are correct or not, man does not go about seeing and thinking in such a manner. The spoken language is a most subtle general instrument to express what man does see and the ideas he does think about. It is a necessary truth that a word or a proper grouping of words expresses either a discriminable *perceptual* figure-ground organization or a discriminable *conceptual* figure-ground organization. This is what distinguishes words from gibberish. When a person understands a word he either perceives or conceives the figure-ground organization it expresses. If he does not understand the word he cannot perceive or conceive of a figure-ground organization which it may express. If he misunderstands the word the figure-ground organization he thinks it expresses is different from that which the user thinks it expresses. It is literally inconceivable for a person to speak meaningfully, to make sense, unless he, and his audience, know the figure-ground organizations which his words express. Hence the paradox experienced by all when we agree that we cannot define 'x' and then proceed to use 'x' in our discussions. Every substantive term or word in any spoken language is a name for a discriminable figure or discriminable part of a figure – the part, of course, can, with the proper shift in figure-ground relationships, become a figure in its own right.

Now, I have argued that figures are anything but the irreducible, unique, unit things that some philosophy seems to consider to be necessary. They are flexible and changing. However, though they admit to being perceived or conceived as having parts or being part of a more inclusive figure, they nevertheless maintain a unit irreducible quality about them as long as they are figures. Substantive words or terms denote this quality. Perceptual figures exhibit the phenomenon of constancy. By this is meant the simple fact that a figure is seen as such despite many noticeable changes in its relationship to the perceiver or conceiver. A white sheet is seen as white whether it be under bright sunlight or dim electric light. So a word maintains a constancy of meaning regardless of the many distinct and different contexts in which it can be meaningfully used. This constancy of meaning which exists throughout the many different contexts or different specific meanings for which the word can be properly used I call the 'core meaning' of the word. When we read a dictionary carefully and critically we see that for words with many specific definitions a common core meaning can easily be identified which differentiates this group of definitions from another group found for another word. Of course, it is sometimes the case that the core meaning is overlaid and hidden by historical accretions and must be uncovered through etymological analysis.

It follows that 'system' has a core meaning, and that that core meaning expresses a perceptual or conceptual figure, a discriminable, distinguishable invariant that can be identified amidst a host of different conditions and circumstances. What is this figure? The difficulty people have in identifying the figure is that they have looked for it in the wrong place; they have looked for it in the objective world, somehow equating the figure expressed by 'system' to be of the same nature as the figure expressed by 'dog', for example. Just as people could easily see the figural invariance of that thing called 'dog', they expected to find a similar invariance 'out there' for that thing called 'system'. But the figure expressed by 'system' is not simply 'out there' – obviously, otherwise there would be no trouble in defining it.

Earlier the terms 'modes of perceptual organization' and 'modes of cognitive organization' were introduced. The use of 'modes' implies that there is more than one mode of cognition and more than one mode of perception. It is my contention that a thing is called a system to identify the unique mode by means of which it is seen. We call a thing a system when we wish to express the fact that the thing is perceived as consisting of a set of elements, of parts, that are interconnected with each other by a discriminable, distinguishable principle. Every one of fifteen definitions of 'system' in my preface is a concrete exemplification of this core meaning.

Whenever one person can point to or explain a set of elements and the nature of the connectivity between these elements to another person, then the other person will perceive/conceive of the set as an entity, a thing. The word 'system' will then spontaneously emerge as the adequate expression, as the proper name for this thing. A system is therefore an interaction between what is 'out there' and how we organize it 'in here'. 'System' denotes an interaction between the objective world and how it is looked at or thought about; it denotes a mode of perceptuo/cognito organization.

'System' is a name for a very general invariance that can admit to very much variation in details. This does not make it fuzzy or ambiguous. 'A solar system' is just as clear and unambiguous as 'our solar system' although the latter case permits knowledge of many details that are in principle unknowable for the former. A general concept, no matter how clear and precisely formulated, tells us little about specific cases or sets of cases that are instances of this concept. 'A solar system' can tell us nothing about the specific number of planets, moons, comets and their respective orbits, that will be found when we become able to see more clearly what goes on around other stars. 'A solar system' does however specify minimum conditions that an aggregate of astronomical entities will have to meet in order to be named 'solar system'. Within the bounds of these conditions there is room for infinite variety.

The mode of perceptuo/cognito functioning which enables us to perceive/cognize systems itself admits to various and different types of more concrete exemplification. There are many different kinds of connectivity which enable man to group entities together to form a system. Hence the fifteen definitions to be found in the dictionary. A step in an attempt to classify and order different types and aspects of connectivity between entities will be made in the remainder of this paper.

A classification into bipolar types that are somewhat similar to dimensions - a possible taxonomy

Structural-functional – static-dynamic. Phenomena can be seen or thought about in two ways. We can attend to those aspects of phenomena which do not change within a defined and delimited time span, and those that do. When the former serve as the perceptuo-cognito figure we speak of structure or of a static stage; when the latter are attended to we speak of function or of a dynamic state. What emerges as a structural figure and what emerges as a functional figure is determined by the time span under attention. If we consider an infinite time span nothing can be structure, as Heraclitus recognized long ago; everything changes in the fullness of time. On the other hand change disappears when we consider an infinitesimal instant of time since change makes no sense except as a specified relationship between at least two distinct instances in time.

The principle underlying the connectivity of a system during a given time period is static if the connection between the entities comprising the system can be seen or understood from knowledge of the state of the system for any one instant within that time period. If at least two instances within the time period are necessary before the principle can be demonstrated, the principle is then dynamic.

Shifting figure-ground relationships apply to systems as to any other kind of figures, hence the same set of entities can be looked at both from a static and a dynamic standpoint during the same time period. For example: the set of space-time points which constitutes a sleeping person during the wee hours of a night can be considered statically – a living organism which is inactive because it is asleep – or it can be considered dynamically – a physiological organism undergoing an anabolism for which sleep is a necessity.

Purposive or non-purposive. So much nonsense has been written about purpose that scientists in general and social scientists in particular have had to, in some sort of self-defense, proscribe the use of the word in scientific discourse. With this, unfortunately, they threw the baby out with the bath, since purposiveness seems to be an essential characteristic of life. By refusing to face 'purpose' the study of life in general, and the study of systems involving life – both physiological and ecological – has become much more difficult.

Like every other meaningful word 'purpose' is a name for a discriminable perceptuo/cognito figure. It denotes a distinguishable pattern of action. What characterizes this pattern is convergence to a terminal state which is called 'its goal'. This convergence seems to be, to a considerable degree, independent of the vicissitudes of the external environment. It is this independence which enables us to assert that living organisms often exhibit great tenacity in achieving their goals, despite the great difficulties the environment confronts them with. In addition, the goal is independent, at least as far as we know, of a point of maximum entropy. Goal-directed action generally decreases entropy rather than the opposite.

The often strange and generally unpredictable shape that the branches of a tree take in a crowded forest is seen/thought-of as the tree's quest for sun. The often equally strange and unpredictable shape the tree's root system will take is seen/thought-of as the tree's quest for water. (Note the word 'quest' which can be used properly in this context – look up its meaning in a dictionary.) The movement of billiard balls exhibits a pattern distinctively different from the movement of basketball players; the pattern of a rock flying through the air is distinctively different from the pattern of a bird. And finally, note how actors have to move in order to communicate to the audience that they are robots or zombies; the goal of the behavior not being the goal of the behaving organism, but the goal of some external power which has the organism in thrall.

Purposive behavior generally can take one of two forms; it can be directed either towards the environment, or towards the system itself. When behavior is directed towards the environment it can either modify the environment so as to create a desired state, or it can overcome difficulties interposed between it and its desired state, or it can seek detours to circumvent and bypass these barriers. In all these cases the specific actions to be taken by the living organism are unpredictable unless one knows beforehand the vicissitudes with which the environment will confront it.

When a beaver builds a dam or a rabbit digs a burrow, we have instances where the environment is being modified. When a dog chews through his leash he is overcoming difficulties interposed between himself and his goal, a state of free movement. When an animal seeks a path through a dense underbrush to reach a source of water which it senses to be on the other side, it is trying to circumvent barriers. It is easy to bring more complex examples of human behavior for all these types of action directed towards the environment but not necessary. These examples from animal life will suffice.

Action directed toward the system itself is omnipresent when a living organism is considered from a physiological standpoint. Cannon, who was the first to stress this aspect of living action, gave it the name of 'homeostasis'. Homeostasis is different from equilibrium. A point of equilibrium in a system is that point where, given the constraints internal and external to the system, entropy is at a maximum. The homeostatic level is anything, as far as we know, but a point of maximum entropy; in fact the organism is almost continuously taking action to decrease its entropy in order to maintain its homeostatic levels.

Physiologically, a living organism can be characterized by constant endeavor to maintain homeostatic balance. Instances of homeostasis are the sugar level of the blood, the body temperature, the water content of the body, the hormonic balance etc. With some generalization of the term, it is also possible to apply it to ecological communities and cultural configurations, since both seem able to initiate action as soon as certain entities that are part of them assume magnitudes above and/or below prescribed levels.

As man's knowledge of his world grew, he found it possible to contrive purposive systems of growing complexity. These systems are characterizable by an input to the system; a processing of the input by the system; and a consequent output which consists of the input as modified by the system. The output of the system is the desired goal which man wished to achieve. In contriving this system man, therefore, had a definite intention in mind. Man-contrived systems are production systems and hence are purposive.

Mechanistic-organismic. Since systems comprise a set of elements or entities, and the connection between elements or entities, it is possible to change, remove, or extirpate elements and/or the connections between elements within them. A system in which the remaining elements, and their connections, undergo no change at this removal or extirpation is perceived as being intrinsically different from a system where they do. In the former case I will call the system 'mechanical'; in the latter case I will call it 'organismic'.

Much of nineteenth-century science shared the ideal that all phenomena are ultimately reducible to a mechanical system consisting of unit elements and a push-pull connectivity between them. With the formulation and development of concepts such as the space-time gravitational field, the sub-atomic electronic field, or chemical equilibrium, this idea has been found to be, most probably, wrong. Natural dynamic systems seem to be organismic. One must almost perforce go along with the conclusion reached by Whitehead that all natural dynamic physical phenomena are organismic. Whitehead, however, seeks a cosmological solution. The aim of this paper is far less ambitious. All that is sought for here is a clarification of how man perceives and thinks about what he calls 'system'.

Man sees many static systems. Geography abounds with them. Mountain systems, archipelagoes, are good examples. If we remove one mountain for fill or connect two islands by a bridge we in no way affect the remaining elements of the system. Most purposive production systems contrived by man are mechanical. So when a machine in a production line breaks down or a part of a machine breaks down no change occurs in the other elements or among the relationships between them. When the broken part of the machine is replaced the system functions as before. Not all static systems are mechanical. One cannot take a part of a soap bubble away nor a part of a suspension bridge. An electromagnetic field can also be considered to be a static organic system.

One can also speak of a partially organic system such that the change, removal, or extirpation will affect a proper sub-set of the system. Damming a tributary to a major river will affect the tributary and the main river below the point of confluence. It will not affect the water flow in the other tributaries or in the main river above the point of confluence.

The emerging taxonomy. The three bipolar 'dimensions' just discussed generate eight cells:

1. Structural, Purposive, Mechanical.
2. Structural, Purposive, Organismic.
3. Structural, Non-purposive, Mechanical.
4. Structural, Non-purposive, Organismic.
5. Functional, Purposive, Mechanical.
6. Functional, Purposive, Organismic.
7. Functional, Non-purposive, Mechanical.
8. Functional, Non-purposive, Organismic.

A road network is easily a good example of a structural, purposive, mechanical system (cell 1). Road maps represent it adequately at a given instant in time, two instants being unnecessary; hence it is a structural system. It has an obvious purpose, that of connecting various communities and other desired geographical points to each other. It is mechanical because one can extirpate any part of it without introducing any change in the remaining parts.

As a structural, purposive, organismic system (cell 2), I will consider a suspension bridge. It is similar to a road network in the first two aspects, but no significant part can be taken from it without disturbing the forces acting upon every part of it. Hence it is an organismic system.

Many examples abound for a structural, non-purposive, mechanical system (cell 3). Let us look at a mountain range. We consider mountain ranges to be systems. They have no purposes, they are just there. If one levels any mountain in the system no conceivable change occurs in the rest of the system within time spans commensurate with a human life span.

Any physical system characterized as being in a static equilibrium can serve as an example for a structural, non-purposive, organismic system (cell 4). Consider an electromagnetic field, or better yet, consider a bubble. Both of these examples can be determined by knowledge of their state at one instant of time. They have no purpose; they just exist. And it is impossible to take any part out of them without changing the entire system, without changing the point of equilibrium.

Functional, purposive, mechanical systems (cell 5), abound around us. Men construct them all the time. A production line is a good example. It makes no sense to think of it except as a temporal succession of steps within which raw material is processed and changed into a desired finished product. It is eminently purposive. If any machine in the line breaks down no change is undergone in any of the other machines in the line even though production may stop. Hence it is mechanical.

Living organisms, *qua* living organisms, are *the* examples of functional, purposive, organismic systems (cell 6). First, what is the meaning of a living organism at an instant in time, in contradistinction to static, structural anatomy? Unless we know its behavior, both internal and external, we do not know it. Behavior is a time-bound process; it is functional. Second, because I have no desire to get into metaphysical arguments as to what is really real and what is really scientific, I will assert dogmatically that the most parsimonious way to understand life at all its levels, from evolution, through physiological functioning, through overt behavior, to cultural and ecological configurations, is by means of purpose. Let us not confuse the mechanisms by which this purpose is achieved and the purpose itself. Perhaps, in some future, purpose can be eliminated and shown to be some sensible function of physical causality. At present this is far from being the case and the stubborn phenomenal facts do show purpose. Third, and finally, an organism is an organism.

Mental elasticity is needed to find an example for a functional, non-purposive, mechanical system (cell 7). Consider the flowing water in a river stream. It is functional since 'flowing' makes no sense unless one takes at least two instances of time into account. Now consider the wild Missouri in its untamed state or that river of tears, the Whang Ho. Both these rivers exhibit a tendency to markedly change their channels occasionally. These changes have been local as compared to the total system, and they have had no effect upon the rest of the system. And rivers, *per se*, have no purpose. Hence the changing flow of water as a result of a change in the river bed can be considered to be an example of a functional, non-purposive, mechanical system.

The last cell (cell 8), a functional, non-purposive, organismic system, will become increasingly important if the physical sciences continue in the direction they have assumed since the formulation of Maxwell's field equations. More and more of the explanations given by the physical sciences to the observable facts of physical behavior are of the nature of a dynamic interdependent field. The atom and the circular four-dimensional space-time continuum are both examples of such a system.

It is interesting to see how the fifteen definitions given in the dictionary fare with this taxonomy. Let us review them one by one.

Definition 1 is similar to the definition of 'system' as given in this paper. But this is hidden, implicit. It is also more limited than the definition presented here since it specifies only two of the six poles identified in this paper: organismic interdependence and function. It mentions four examples. The first two, the universe and the solar system, clearly belong to cell 8. The next example, a telegraph system, is contextually an example of cell 5. We can, however, consider it from a structural standpoint exclusively, and then it belongs to cell 1. The last example, a quote from literature, a system of hills, is clearly equivalent to my example for cell 1, a mountain range.

Definition 2 includes the universe again, which we have already treated. It also includes the body from a functional physiological standpoint, and from a psychological standpoint. Both of these fit into cell 6.

Definitions 3 and 4 both deal with the same class of things, systems of ideas connected to each other by some rational or coherent principle. These systems are structural and mechanical, and (somewhat elastically) I believe they are non-purposive, though, of course, the person who constructed them did have a purpose in mind. Hence they are examples of cell 3.

Definitions 5 and 6 deal with verbal instructions or formalized models of procedures. They are structural, purposive – they exist in order to instruct a person what to do, and they are mechanical. Hence they belong in cell 3.

Definition 7 concerns 'the system'. It is colloquial American and concerns the interests and powers which control the government to a greater and lesser extent. This system seems to be functional, since 'control' is time-bound, purposive, and mechanical; a denizen of cell 5.

Definition 8 is about physiological sub-systems that contribute toward vital physiological functions of the organisms. It obviously belongs to cell 6.

Definition 9 is a very interesting usage of the word. It names a body of evidence submissible to a court which points to the intention of the defendant. Since 'pointing to' is one of the perceptuo/conceptuo criteria for purpose, this system is purposive. Since it exists at a given instant in time it is structural. It is also organismic and interdependent, as it is the *totality* of the evidence which points; its meaning can change with the exclusion of any specific bit of evidence. Hence it belongs to cell 2, the only definition that fits into this cell.

Definition 10 is about a division of rocks in geology. Such a division is structural. It just happens to be there and has no purpose. In addition, it is mechanical. Obviously a candidate for cell 3.

Definition 11 concerns an identified poetic form, structure. It is structural, non-purposive, mechanical and belongs to cell 3.

Definition 12 concerns the way 'system' is used in music and is conceptually very similar to definition 11. However, since music is time-bound, musical systems are functional. It is therefore a member of cell 7. The only definition to fit that cell.

Definition 13 concerns chemical systems in dynamic equilibrium. These have already been discussed as members of cell 8.

Definition 14 concerns a transportation system. It is discussed from a static standpoint, e.g. a large group of lines under common ownership. Since it is also purposive and mechanical it belongs to cell 1.

Definition 15 deals with physiological systems from a static, anatomical standpoint. As such it becomes static, non-purposive, and mechanical and belongs to cell 3.

Every one of the definitions seems to fit, without too much conceptual violence, into one of the cells. It is very interesting to note that not one of the systems defined in the dictionary fits into cell 5, a functional, purposive, mechanical system for which the example given was a production line. Other usages of 'system' that are common nowadays, like man-machine system, command-control system, and weapons system also seem basically to belong to this cell. The eight cells were generated through a systematic analysis of what the meaning of 'system' must be to make sense. Dictionary meanings are obtained through a thorough and assiduous 'nose counting'. The fact that some of the best lexicographers in the country missed the specific meaning of 'system' in cell 5 seems to point, at least to me, to the superiority of even a simple intelligent analysis over the most elaborate and careful 'nose counting' and classification of what was counted.

In addition, it is interesting to note the example of the use of 'system' quoted from Owen Wister: knotted systems of steep small hills. This usage does not conform to the general definition given just above it. Hills do not interact nor are they inter-dependent. They do not form an integrated, organic, or organized whole. Nor do they function, operate, or move in unison. They are grouped together and perceived as a system on the basis of similarity and proximity. But the general definition cannot handle this. The systematic analysis can and does.

Some other ways of thinking about systems – not dimensions

Self-organizing systems. The term 'self-organizing systems' is basically an instance of verbal magic that accompanies the changing of a name. As a result of exigencies of the Second World War, a technological breakthrough occurred in the design of control systems; simple control mechanisms have been known for centuries. The new control systems could sense subtle changes in the environment and could, as a result, modify the functional purposive system which they were designed to control. These modifications generally changed the function of the system in light of changes in the environment. This is where feedback came into the general vocabulary. Feedback was part of the electronic designer's vocabulary in the years prior to the war and, as already mentioned, was known for a longer period of time. What was radical in the new development was that now, for the first time, man developed a physical system which could 'see' the external world and change its behavior in a manner appropriate to what was 'seen'. The analogy with life sprang forth immediately and was met with excitement and enthusiasm – at last a way was found to base living phenomena upon a physical substratum. The control systems were supposed to be analogous to the brain, and the systems they controlled, to the body. The total system was then given a name which was immediately, explicitly and/or implicitly, applied to living organisms.

In the ensuing enthusiasm the very serious deficiencies of the analogy were not attended to. I can list many but will concentrate upon only one, since that is all that is necessary. It is an uncomfortable fact that man sees objects that are far away from him, that he sees at a distance. The senses of the control system cannot; they can react only to the stimuli impinging upon their sense organs. Now it is true that man cannot see at a distance unless there are physical stimuli impinging upon his sensorium too. These are called proximal stimuli. But what he actually sees has a very tenuous relationship to the proximal stimuli. The only thing a control system can react to however is the proximal stimulus distribution. Psychologists denote this ability to see 'through' the proximal stimulus to the distant object by the term 'vicarious mediation'. Control systems do not exhibit any ability to respond to mediation vicariously. Hence the analogy is not well taken.

The conceptual confusion accompanying 'self-organizing system' results from the fact that explicitly and/or implicitly the scientists in defining this term wish to include both a certain set of physical systems and living organisms. This cannot be done since the set they have in mind is only superficially similar to living organisms and, although they may resist recognizing this, they feel it in their guts. If we restrict ourselves to physical systems the term offers no difficulty. Most systems contrived by man function mechanically in a prescribed manner. To change their functioning the intervention of a human operator is generally needed. To the extent that we can construct control systems that can change the functioning of the system without human intervention, to that extent we have self-organizing systems. It's as simple as that.

Central and peripheral properties and/or elements of organismic systems. In the preceding section it was noted that the self-organizing physical system was superficially similar to human organism. In other words the aspects in which it was similar to a living organism were not important aspects. The ability to react to physical stimuli is not, *per se*, an important aspect of living/ functioning; it is the ability to react to a distant object over a wide range of mediating stimuli which is important and which differentiates life, at least many advanced forms of life, from sensory machines. This points to a problem which can be generalized, a problem which does not seem to exist for mechanical systems. The problem can be crudely formulated as 'When is a man a man?' or 'When is a solar system a solar system?'.

Aristotle touched upon this problem when he asserted that a hand separated from a body is not a hand. I am not aware of him considering that a person without a hand is still a person. A hand, therefore, although it itself has no meaning as such unless attached to a person, is peripheral to the person as far as him being a person is concerned. In fact we can subtract all kinds of things from a person without him ceasing to be a person. Wars generally contribute to human progress in many scientific fields. Medicine has learned much in the last war. One of the things it learned is how, more efficienctly, to keep people alive despite all kind of fantastic external dismemberment which is never met with in peacetime. Hence we now have a small number of quadruple amputees living among us. They are still perceived as men. Limbs therefore are peripheral characteristics of human beings. Surprising? No! Monkeys have four human-like limbs.

On the other hand, idiots, neonates, or psychotics are generally not perceived as persons even though their bodies are intact. Rational consciousness and the behavior flowing from rational consciousness is a central property of man. We find something similar on a physiological level. The body can maintain relatively efficient life processes with many of its parts being subtracted from it. But there are other parts that are so essential to any ordered function that their slightest damage will cause death. Hence we have parts that are central for a physiological life and others which are peripheral. If we go to a solar system we find something similar but far simpler. In order for a solar system to be perceived/cognized there must be a sun and at least one planet circling it. All other possible aspects and properties of a solar system are peripheral.

The most general definition, the core meaning, of an organismic system must be restricted to its central properties. The inclusion of peripheral properties will almost always exclude certain instances of this organismic system which lack this peripheral property, but are still seen as being the same as the defined system. This introduces confusion. A conceptual definition should be in accord with that which we see or think of spontaneously. I find it difficult to think through the problems entailed in how we perceive and discriminate centrality from peripherality in a systematic way. Nevertheless we do discriminate and if we disregard these discriminations we can often run into conceptual difficulties. One will learn very little about an organismic system if he focuses his attention on its peripheral aspects.

Conclusion

The difficulty in defining 'system' in a specific context results from misusing a word which has a simple, clear meaning in a general context for a specific, concrete context. It is similar to the classical fallacy denoted by *pars pro toto* or that which was called by Whitehead 'the fallacy of misplaced concreteness'. 'System' is at a level of generality similar to 'phylum'. If we know the phylum to which an organism belongs and nothing else we know very little about the organism. Ditto for system. The only things that need be common to all systems are identifiable entities and identifiable connections between them. In all other ways systems can vary unlimitedly. The quest for a more detailed, specified definition for 'system' is chimerical. The same holds for a quest for a general system analysis.

However, as I have attempted to show above, it is possible to group systems according to specifiable characteristics. The definitions for systems belonging to such groups become more detailed and specific. It is a fact that for many such groups, at present ill-defined, if defined at all, certain analytic techniques are very appropriate. But it does not follow that they are appropriate for other groups of systems. By recognizing this we know better where we stand, the air gets clearer and we can see more clearly. And with this we can think better.

4 Joseph E. McGrath, Peter G. Nordlie, W. S. Vaughan, Jr

A Descriptive Framework for Comparison of System Research Methods

Excerpt from J. E. McGrath, P. G. Nordlie and W. S. Vaughan, Jr, *A Systematic Framework for Comparison of System Research Methods'*, Human Sciences Research, Inc., 1960, pp. 13–36.

The heterogeneous content of system research studies, and the equally heterogeneous array of methods which they have employed, have produced a bewildering collection of terms and concepts descriptive of different aspects of system research methodology. Effective classification and comparison of methods requires a sorting and simplification of these terms into a few basic, integrative concepts which can serve as anchors for further developments.

Several such key anchoring concepts, and further methodological distinctions which derive from them, have emerged from review of reports of system research studies and discussions of system research methodology. These concepts provide a general framework within which various methods and techniques employed in system research studies can be classified and compared. This general methodological framework, considered as a first approximation, is described in the present chapter.

Structure of a system research problem

Any system to be studied usually is a part of a larger system and is composed of more molecular systems. The entity defined as the total system for one study may be a portion of the defined system of another study, and a supra-system with respect to the defined object of still another system study. The present research accepts each investigator's defined object of study as the total system for that study.

A system research study is defined as a research effort whose purpose is to contribute to development of an optimal man-machine complex. Such a study may relate to an entire system, or part of a system; it may contribute to design or development of a 'new' system, or to evaluation and improvement of an existing system. Thus, a wide variety of specific research problems fall within this definition.

There are at least three major aspects of any system research problem:

1. The system itself, and its parts.
2. The performance of the system in relation to its purposes or objectives.
3. The system's environment – the medium in which it is embedded and in which it operates.

Each of these aspects of the system research problem can be described in terms of a number of properties or characteristics. In the present discussion, such properties will be called variables.[1] *Research on a system requires obtaining information about relevant variables from each of these three aspects of the system research problem.* A property or variable descriptive of the system, of its medium, or of system performance, is *relevant* to a system research problem to the extent that differences in that variable directly or indirectly affect the degree to which (and the efficiency with which) the system accomplishes its objectives. Types of variables from each of these three aspects of a system research problem are discussed next, followed by a discussion of the basis of relevance of a variable to a system research problem.

System variables

System variables refer to properties descriptive of the entity which is the defined object of study. They include structural and organizational properties, informational characteristics, and properties descriptive of the operations by which the system acts. The goal of design or modification of a system, at the most general level, is to fix all relevant system properties at values which provide optimal levels of system performance.

System variables can be descriptive of various levels of organization of the defined system. For present purposes, three levels of organization of a system will be distinguished.[2]

1. *Total system:* the man-machine complex which is defined as the object of study, as bounded by the investigator;

1. It is noted that the term *variable* has been used with several different meanings in the system research literature. The term *variable* here refers to any property of an entity which can assume different quantitative or qualitative values on different occasions.
2. It may be useful for various purposes to define and deal with four or more levels of organization. Three levels are dealt with here because, from the viewpoint of the present study, the same types of research activities would apply to the additional levels.

2. *Major sub-system:* proper parts of the system, which may be distinguished on a structural or a functional basis;

3. *System components:* the constituent parts – human or equipment – of a sub-system.

System performance variables

System performance variables refer to properties descriptive of the *results* of operation of the system, with respect to attainment of the system's goals. Performance variables also can be descriptive of different levels of organization. Again, three levels are distinguished for classification.

1. *Objectives:* the overall purposes of the system.

2. *Functions:* major classes of actions or performance required to achieve the objectives.

3. *Performance requirements:* specific dimensions describing required actions and a statement of a standard or required level of performance on each.

Levels of system variables, described previously, do *not* map one to one with levels of system performance variables. Several sub-systems may contribute to performance of a single function, and a given sub-system may contribute to performance of several functions. Further, the performance of any system component can be described in terms of a number of performance requirement dimensions.

Environment variables

Environment variables refer to properties of the system's surroundings which impinge on and affect the system and its performance. What is system and what is environment, in any given case, is often arbitrary. For purposes of this study, the bounds by which an investigator defines his object of study also define what is environment.

Environment variables include properties of objects and of physical and less tangible forces. For weapons systems, the relevant environment often includes the capabilities, actions and intentions of an enemy as well as properties of objects and forces in the physical and socio-cultural medium in which the system functions.

Relevance of a variable

The basis of relevance of a variable is the extent to which differences in it are associated with differences in degree of attainment of system objectives. Variables descriptive of the system itself can affect the attainment of system objectives by affecting one or several aspects of system performance. Variables descriptive of conditions in the system's environment can be relevant either because they affect the system directly, and thus alter its performance (e.g. low temperature preventing a weapon from firing), or because they affect system performance and thus alter the degree of fulfillment of objectives (e.g. wind deflecting the path of an accurately aimed projectile). A variable descriptive of system performance can affect attainment of system objectives directly (i.e. it may be descriptive of the level of performance of one particular aspect of the objective) or it may interactively affect other system performance variables (e.g. high speed in an aircraft may *preclude* high maneuverability).

Basically, then, the research information required about a system research problem consists of: (1) specification of all relevant variables of system, environment, and system performance; and (2) delineation of the basis and extent of their relevance – the relationships which express the effects of their variation on the attainment of the system's objectives.[3] The manner in which these two classes of research information are obtained and utilized in various kinds of system research investigations is discussed in the next section, in the context of a descriptive classification of the process of system research.

Descriptive classification of the system research process

In concept, conduct of a 'complete' system research investigation requires specification of all relevant system, system performance and environmental variables; determination of the existence, degree and form of relationships among them; and utilization of that information to construct a system which is optimal with respect to its objectives. In practice, conduct of a complete system research investigation, even for a relatively simple system, is usually not feasible within the confines of a single study. Characteristically, a given system research study has a more limited

3. Statements of the effects of variation in one variable on the occurrence of values of a second variable are considered *first-order relationships*. Expressions of the interdependence among three or more variables are termed *higher-order relationships*.

objective and emphasizes certain portions of the overall system research process as defined above.

From a methodological viewpoint, the overall system research process can be structured into more homogeneous sub-units in terms of two classification dimensions: research stages, which reflect different study purposes, and research functions, which refer to different types of research activities. These two classifications of the system research process are discussed below.

Research stages

Characteristically, a given system research study is oriented toward one or another of a series of purposes which are inherent in the development cycle of a system. Four basic purposes, or stages, of a system research problem are identifiable.

1. Delineation of system performance requirements.
2. Derivation of design consequences of requirements.
3. Development and integration of the system.
4. Evaluation of system performance in terms of requirements.

These four stages form a closed-loop cycle which parallels the development cycle of the system. They contribute in different ways to obtaining the required research information discussed at the end of the preceding section. At the initial stage of development of a system, there is a need for specification of the variables descriptive of system performance, and required or desired levels on those variables. Then, hypotheses as to design features which will achieve the required levels of performance are needed; these imply that certain system variables are relevant (i.e. that they affect system performance), and that certain values on them will lead to optimal system performance. Development and integration of the system itself provides a physical expression of these hypotheses. Finally, evaluation of the system's actual performance in relation to its objectives is a validation of the correctness of the hypothesized or inferred relationships.

Research functions

The research stages are methodological units of system research in the sense that they define domains of effort and purpose within a total system research problem. Within each of these stages, there are different kinds of activities or functions by which research information is obtained and utilized.

Three different types of research functions contribute to the system research process: (1) development of models, (2) collection of research information and (3) synthesis of information within the models. These functions are mutually dependent. No model of any portion of a system research problem can be meaningful or useful unless it is adequately grounded in relevant information about the system problem under study – the 'real world'. On the other hand, activities aimed at generating relevant information about the system problem literally cannot be carried out unless guided implicitly or explicitly by some model or structure; and resulting information is not meaningful until it has been synthesized and organized in some fashion.

Models follow from and elaborate the statement of the system research problem, and serve to guide information collection and information synthesis activities. Reciprocally, research information about the system problem guides initial development and later evaluation and modification of the model.

While all three research functions are carried out in each stage of the system research process, their relative importance varies for different stages (see Table 1). Stage one puts emphasis on development of a research model. Stage two emphasizes synthesis and interpretation of information. Stage three emphasizes

Table 1 **System research stages and research functions**

	Research stages			
Research functions	Requirement setting	Delineation of design consequences	Development and integration of systems	System evaluation
A Model development	Emphasis on development of research model	Provides criteria to guide C	Emphasis on development of design model	Provides criteria for interpretation of B
B Information collection	Provides information instrumental to A	Provides information instrumental to C	Provides information instrumental to A	Emphasis on generation of information about system performance
C Information synthesis	Integrates information instrumental to A	Emphasis on interpretation of design consequences of requirements	Interprets information to guide A	Emphasis on interpretation of B, with respect to requirements

development of design models. Stage four emphasizes collection and synthesis of information.

The following sections of this chapter discuss each of these three research functions, in turn, developing further classification distinctions within each. These classification distinctions provide categories for description and comparison of methods and techniques used in system research.

Development of models in system research

The term *model* is here used broadly to refer to any representation of all or a part of a system research problem. Although the term has been widely used in system research literature, it does not have a single, clear and unequivocal meaning. A variety of theoretical structures, serving a variety of methodological functions in the research process, have been designated as models.

Types of models

Basically there are two kinds of models; models descriptive of the system itself, which are termed *design models*; and models descriptive of variables and relationships relevant to system performance, which are termed *research models*.

Design models vary in comprehensiveness, and in the degree to which they represent the system in concrete and dynamic form. Initially, design models may be formulated in verbal terms. Later, design concepts may be expressed in functional diagram form, or as blueprints. Eventually, design models become physical representations of the system (breadboard models, soft and hard mockups, etc.). Design models become more complete, concrete and dynamic as system development progresses, and ultimately become prototypes of the actual system.

Research models are formal statements of the variables presumed to be relevant to a system research problem, and of the manner in which they are presumed to be related to one another and to the attainment of the system's objectives. A research model can be considered either as an efficient summary of empirical knowledge about the problem, or as an interrelated set of hypotheses to be validated by application of information gathering and information synthesis methods.

A research model may be expressed in qualitative (verbal descriptive) or in quantitative (mathematical) terms. The important distinction between models is not whether words or mathematical symbols are utilized (anything that can be expressed in mathematical symbols can also be expressed in words) but the extent to which terms and relationships among terms are expressed in *precise form*.

Generally, verbal descriptive models are stated in less precise form, and thus presume less complete prior knowledge about the nature and form of relevant variables and relationships. A mathematical model derives its power not from the use of symbols as such, but from the relative precision and detail with which it can represent relevant variables and their interrelationships. Mathematical models, therefore, usually presume a more advanced level of knowledge about variables and relationships.

Development of research models

Initial formulation of a research model is often in the form of a generic statement of relationships. If it is a mathematical model, the generic statement might be: 'Y (the over-all criterion o system success) is some function of X_1, X_2, . . . X_i, . . . X_n (relevant variables).' The model-development process then becomes one of more and more precise formulation of that generic function, by elaboration of the ways in which the elements of the problem are related to one another. This process is presumably guided by existing empirical information, and may at times require the launching of information gathering activities to determine specific key portions of the model. In the main, however, model development is essentially a process of simplification. Model development tends to reduce the complexity of the statement of the problem by excluding marginally relevant variables and by translating empirical information about relationships into more simplified forms so that the model will be more amenable to solution by existing computational techniques.

Models for recurring problems

At a general level of description, widely differing systems exhibit many similarities in process. System research studies often take advantage of such similarities by utilizing general mathematical models which have been developed for certain recurring problems. Churchman, Ackoff and Arnoff (1957) have discussed and classified some of the types of focal problems for which mathematical models have been developed. An adaptation of their classification of major system processes, the principal types of recurring problems for which mathematical models have been developed,

Table 2 Classification of models by type of focal problem

Type of system process	Type of focal problem	Form of problem	Mathematical synthesis techniques
Planning	Inventory (production-inventory)	Economic lot size, order-time	Various analytic and iterative techniques
		Restrictions on capacity	Linear programming techniques
Programming	Allocation	Distribution assignment	Linear and non-linear programming methods
Scheduling	Waiting line (Queing)	Determining facilities	Analytic, iterative, and Monte Carlo methods
		Scheduling arrivals	
	Sequencing	Line-balancing Routing	
Maintenance	Replacement	Degenerating units	Dynamic programming methods; various iterative techniques; Monte Carlo methods
		Go/no-go units	
Strategy	Competitive	2-sided games	Linear programming techniques; various iterative techniques; Monte Carlo methods.
		Bidding	

and some of the mathematical techniques employed for synthesis of information within such models are presented in Table 2.[4]

4. The mathematical techniques used for solution of the model are listed in the right hand column of Table 2 and can be discussed as methods for synthesis of information in the next section of this chapter. The table scheme of classification is based upon C. W. Churchman *et al.* (1957).

These focal problems, and the types of models associated with them, are not related in a simple fashion. For example, Churchman *et al.* (1957) point out that inventory problems with certain types of restrictions shade into allocation problems; that replacement problems can be viewed as inventory problems, as allocation problems, or as competitive games.

In concept, a total system is likely to include all or nearly all of the system processes listed in Table 2; hence, a total system problem is likely to include all of the types of focal problems listed. In practice, a given research study is apt to develop a model which gives emphasis to one or another of the recurring problems. Selection of which problem to emphasize, and therefore of which type of model to develop, is likely determined in part by the nature of the system being studied, in part by the level of development of the problem at which a given study begins, in part by the existing mathematical and methodological state-of-the-art, and in part by the experience, training and biases of the investigating team.

Collection of information in system research

Information collection in system research consists of:

1. Gathering or generating knowledge about the variables relevant to the problem and their range of variation.

2. Gathering or generating knowledge about relationships – systematic covariations of two or more variables – which set restrictions on the combinations of values of relevant variables that can actually occur as the system operates.

Information sources

Information about variables and/or relationships within a system research problem can be obtained from any one (or a combination) of several kinds of information sources:

1. Logical assumption.
2. Indirect empirical sources: (system experts, system users).
3. Direct empirical sources: (results of controlled observation or experimentation).

These information sources are listed in increasing order in terms of the degree to which they contain internal mechanisms for *estimating the probable error* of the information which they generate, and therefore in terms of the degree of confidence with which that information can be employed. However, it should be recognized that this is *not* an order of unqualified goodness, for the more powerful empirical methods are also more costly and time consuming, and are usually accompanied by marked restriction in generality.

In concept, any one of these sources can be used to obtain any or all of the research information required for a system research problem. In practice, however, certain sources appear to be more useful than others for obtaining certain types of required information. For example, it is probably relatively ineffective to try to get precise quantitative information about the distribution of variables, or about the exact functional forms of relationships between variables, from reports of system experts or users. More often than not, such precise specifications are obtained either by controlled experimentation or by logical (mathematical) assumption. Conversely, it could often be very difficult to use direct empirical means to make an initial search for variables relevant to the problem; while interrogation of system experts or users often provides an effective means for obtaining such information.

Direct empirical methods

Methods which can be used for gathering information via direct empirical sources can be classified along a dual continuum of increasing research control and decreasing operational realism. Along that dual continuum several major categories of methods, common within research literature, can be distinguished:

1. Studies in Operational Settings: Information collection from observation of the actual operating system, with a minimum of interference with ongoing operations.

2. Field Experimentation: Information collection from observation of the actual operating system but with major experimental interference with ongoing operations – sometimes by restrictions of the scope and range of environmental conditions, sometimes by restriction of the range of system tasks which are included.

3. Laboratory Experimentation: Collection of information about limited portions of the operating system, while major portions of that system are represented under close experimental control. This can take either of two forms:

(a) *Simulator studies*, in which major portions of the system are represented as realistically as possible, but under close experimental control;

(b) *Laboratory experiments*, in which major portions of the system are represented in relatively abstract form.

As research control is increased, relationships can be specified with greater precision because more and more of the relevant variables are brought under control of the investigator. At the same time, however (assuming that many of the variables being controlled in experimental settings are *not* studied over their full range of variation), results obtained from such controlled information gathering situations have less generality than results of studies done in the operational situation itself. Thus, the greater the degree of experimental control, the more precision but the less generality of resulting information. Conversely, the less the experimental control, the greater the realism but the less easy it is for the investigator to precisely pin down the research information which is contained in his results.

Use of the more controlled methods implies greater prior knowledge of variables and relationships among variables within the system problem. For example, use of field experimentation implies knowledge of what variables are relevant, and therefore need to be controlled and/or measured, while it provides a means for determining relationships between those variables. Obviously, such information gathering methods are subject to confounding effects of all 'unsuspected' but relevant variables – *i.e. they are subject to errors contained in the information which their use presumes*. Similarly, use of laboratory simulation implies more precise knowledge of variables, of some relationships, and even of some higher-order interactions which are built into the simulation system. At the same time simulation permits determination of higher-order effects which normally can not be studied in less controlled information gathering situations.

Synthesis of information in system research

Information can be synthesized so as to provide integrated knowledge bearing on the system research problem, by use of any of a range of computational methods. Techniques for synthesis of information provide the basic mechanism by which detailed information is translated into information at higher levels of the system problem.

Mathematical synthesis techniques

Mathematical techniques for information synthesis range along a dual continuum of decreasing specificity of the solutions which they provide, and conversely, increasing complexity of the problem formulations to which they can be applied. Along this continuum, three major classes of information synthesis techniques can be identified:[5]

1. Closed analytic solutions: e.g. solution to n linear simultaneous equations in n unknowns; solution to quadratic equation in 1 unknown, differentiation and integration of certain functions.

2. Iterative approximation techniques: e.g. numerical analysis techniques such as the Newton-Raphson iteration, factor analysis methods.

3. Stochastic estimates: Monte Carlo techniques.

The synthesis techniques which provide more definitive forms of solution are likely to deal with less comprehensive formulations of the system problem. Conversely, the less specific the solutions provided by a type of information synthesis method, the more likely that it can handle problems which include many relevant variables and relatively complex forms of relationships.

Synthesis techniques and mathematical models

The formal properties of the model guiding the collection of information, and the type of information gathering method used, place constraints on but do not entirely fix the type of mathematical techniques that can be used to synthesize resulting information. The mathematical form of a guiding model (stochastic/deterministic) is not necessarily preserved in the data synthesis methods which are used in its solution. Analytic techniques and iterative techniques can be utilized for both deterministic and stochastic problems. Monte Carlo methods, principally used for solution of complex stochastic problems, can be used for approximate solutions to deterministic problems. On the other hand, initially stochastic formulations are often converted into 'expected value' deterministic formulae to permit solutions.

It is also clear, by reference to Table 2, that there is no simple, one-to-one relation between the classification of mathematical

5. The three-way classification used here follows Churchman, *et al.* (1957). Goode (1957) has listed four model types, cross-classified on the dichotomies: Deterministic *vs.* Stochastic, and Analytic *vs.* Numerical. In the context of the present discussion, the formal characteristics of mathematical models as formulations are represented by the Deterministic *vs.* Stochastic distinction. Goode's Analytic *vs.* Numerical distinction is here considered as descriptive of techniques for *synthesis of information.*

models by focal problem types, and the mathematical synthesis techniques which have been employed in their solution. For example, linear programming techniques have been used extensively for solving allocation problems – in fact, they are almost defined as techniques for providing optimal allocations. But they have also been employed in what would normally be characterized as inventory and maintenance problems, and the similarity between linear programming and some game theory problems has been established.

Utilization of results of synthesis

Information synthesis techniques provide the mechanism for combining sets of research information into integrated statements of the relevant variables and their interrelationships. If all possible events relevant to the system problem are properly designated, and all restrictions in those events due to relationships among variables are determined, the set of possible combinations of values which result from information synthesis then constitutes the matrix of possible outcomes that can occur in operation of the system – the remaining uncertainty in the problem. This matrix of possible outcomes may or may not indicate a unique, 'optimal solution' of the research problem. Rather the matrix of possible outcomes may indicate that a number of design variants lie in a 'zone of indifference' such that the set of possible outcomes associated with each of several design variants appears to be about equally favorable for success of the system. When this occurs, additional criteria are needed to fix system design. In this context, the goal of system design (or modification) is to insure that relevant variables descriptive of the system itself are set at the combination of values which is associated with the most favorable set of possible outcomes when the system is operating in its intended environment.

References

CHURCHMAN, C.W., ACKOFF, R.L., and ARNOFF, E. L. (1957), *Introduction to Operations Research*, J. Wiley.
GOODE, H.H. (1957), 'The application of a high-speed computer to the definition and solution of the vehicular traffic problem', *Ops. Res.*, no. 5, pp. 775–93.
GOODE, H.H. (1957), *System Engineering: An Introduction to the Design of Large-Scale Systems*, McGraw-Hill.

5 George O. Wright

A General Procedure for Systems Study

Excerpt from G. O. Wright, *A General Procedure for Systems Study*,
Wright-Patterson Air Force Base, Ohio, 1960, pp. 1–13.

Knowledge about the nature of the machine complex in which
man is expected to work is a necessary background for understanding systems functions, because the nature of the machines in
part determines the requirements of human performance. It is
fairly obvious, however, that analytical schemes focusing exclusively on the characteristics of the machines offer only restricted
insights into system performance. Since the man-machine system
is in fact a complex of interactions, direction of study must
include the man as well as the machine.

How then can we study systems so as to take account of this
interaction? The thesis here proposed is to consider the man-
machine complex as a cultural system. Human culture is defined
as the 'patterned and functionally interrelated customs common
to specifiable human beings . . . and social groups' (Gillen, 1948).
Machines as artifacts of the culture are not specifically included in
this definition of culture; however, they are either the products of
these patterned activities or adjuncts of them.

Let us, therefore, talk about systems study in cultural terms.
This is justified because no machine has any functional independence from human culture. The weapons of war, for example,
are meaningful only in reference to humans and their patterned
activities. The whole case for systems study is not so simply
described, however. We cannot gainsay the fact that at some
point in systems study it will probably be appropriate to shift
our orientation from a study of culture to technique. To illustrate:
we may find we can set the boundary conditions for the design of
a radar set using the theory and methods of a study of culture;
yet once we have set these boundaries, it will be appropriate to
design the radar set in accordance with engineering principles.
That the influence of culture is accounted for in setting these
system boundaries is the particular thesis of the present paper.

What we choose to focus upon in the study of systems is

largely a matter of analytic convenience. The vocabulary we use is, of course, a consequence of that choice. If we focus on machines, the language is likely to be that of the engineer. If we focus on the human, we are likely to use the language of the social scientist. If, however, we wish in systems research to provide a common meeting ground for both the social scientist and the engineer, we must choose a language that is itself neutral. In other words, the language we use must be suitable for communicating the ideas of the social scientist, let us say, to the engineer. We must be able to transmit to him statements about the findings of the social scientist in such a way that the engineer can then act appropriately. A neutral vocabulary would serve to bridge the gap between specialists in disciplines that have their own peculiar vocabulary. It is to serve this objective that our first model is cast in a neutral framework. This so-called neutral model is developed in accordance with the following criteria:

1. It should create a vocabulary that permits all of the elements of the system to be described in common terms. Both man and machines must be described in accordance with some common referents, such as *process* or *link position*.

2. It should use terms and concepts that are subject to experimental manipulation. The scheme for analysis should be in terms of variables and their interrelations.

3. It should be consistent with the pertinent body of psychological and engineering knowledge. The consequence of this criterion is that, in effect, established scientific information can be brought to bear on the systems problem. This does not imply that new principles cannot be developed, but rather that no new principles should contravene established scientific findings.

4. It should lead to the creation of theoretical principles that explain and predict system performance. A logical consistency must run through the entire scheme and, moreover, implicit to the generation of principles is the necessity for the experimental verification of their accord with real life.

The model: neutral framework

In order to create a model of system behavior, we need a concept to describe all of the elements in a common framework. The concept we have chosen is *process*. This frees us from the necessity of describing machine performance and human performance in different terms. In fact, the process description is independent of the entity performing the process. For example, we may describe data gathering or data transformation as processes, and yet they can be performed either by men or machines.

At one level of generality we can characterize system performance in three domains. First, there is the *input* of data. These data may be information about terrain or the location of enemy targets, and may include also information about the location of one's own airplane in space. The essential character of these data is that they will, or can be made to, enter the system. Second, there is the domain of *system processing*. Substantially, this domain embraces all of the distinguishable things that can be done with and to the data. Generally speaking, we may collect the data, we may transform it, we may translate it or cause it to flow through the system, and we may use it to control operations. A special case of processing is the storage within the system of input data. Finally, there is the domain of *system output*. System output generally includes the performance of the task or subtask elements. We may, therefore, characterize a system in terms of its input, process and output, including feedback, if it exists.

We shall attempt now to enlarge the paradigm of input, processing and output. The enlargement that is sought is one in which additional details can be derived that conform to the original four criteria. The reader is referred to Figure 1 for a schematic representation of the relations among the elements that will be discussed.

The first additional element of the model deals with data *transformation*. The data transformation function in the system model provides for the modification of the data into a form that is usable in the system generally. For example, data about the target terrain for a radar bombardier has to be transformed in the linear dimension and also has to be changed from three dimensions to two. The expanse of terrain may cover 200 miles in real space; it has to be reduced to the dimensions of, for example, a seven-inch oscilloscope. Some oscilloscopes, moreover, can accommodate only two dimensions; consequently three-dimensional space has to be transformed for such a presentation.

In general, then, if we are given a set, S, of elements, ι, then the transformation γ of S is a rule which makes correspondence of each ι of S a unique element, ι' of S. Stated another way, $\iota' = \iota\gamma$, and we say that ι' is the γ transform of ι. The data transformation function operates in accordance with this rule.

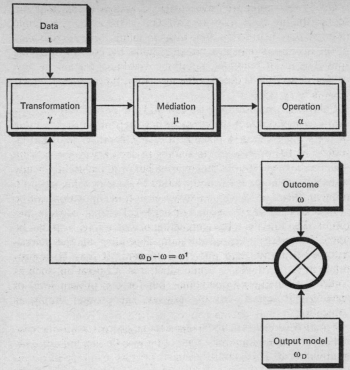

Figure 1 The systems model

Another element is *data mediation*, the operation whereby meaning is added to the transformed signal inputs. We shall consider that the mediation process has various weighting and sorting functions built into it. For example, let us suppose that we are free to look into the transformation process. There we find a signal about an aircraft target. It is necessary to add meaning to this signal from the system's point of view. Is the target friendly or hostile? If it is hostile, how fast is it moving? The mediation function we are describing brings the built-in functional process to bear on the signal and provides the system with answers to questions of meaning. The data output from the mediation process is tagged with notes about size, speed, hostility and estimates of future performance. We consider that the data output at this point also is addressed to the appropriate addressee at subsequent places in the system.

Up to this point we have created a system model that describes the life cycle, as it were, of data in the system. A simple example may help to fix these ideas in mind. Suppose that part of the operator's task in a system requires the control of engine rpm. The meter may have additional markings for *low* or *high* or *danger* serving as cues for the mediation function that directs what is to be done.

Following this example, we can add the next functional element, *operation*. Once the operator has determined the meaning of the data, he is in a position to act. Operation shall be described as the process of responding in such a way as to bring about a change of state in the environment or to maintain a steady state. The operator may choose either to do something or to do nothing. Whether he chooses to act or not, his operation can be evaluated in terms of the kind or degree of change of state that results. The focus can be upon acting or not acting, or it can be upon the resulting change of state, depending on the investigator's orientation and analytical concern. It may be appropriate in some studies to define subclasses of operation, such as control or emergency operations, but for our present level of generality it is sufficient to designate the process simply as 'operation.'

Four additional functional elements or processes will increase the convenience of analysis. One of these is designated *outcome*, symbolized ω. This class of element permits arbitrary measurement categories to be established. We might, for example, create such subclasses of outcome as success or failure. Another additional functional element is designated *output model*, symbolized ω_D. This element includes all of the arbitrary criteria of systems performance by which outcomes can be evaluated. It is by reference to the output model that acceptable performance can be determined.

The description of two universal processes completes the functional model. First, that of feedback control: in this process, the failure of correspondence between the output model and the outcome is determined. This difference, $\omega_D - \omega = \omega'$, is called the *system error* and is fed back into the system for the control of performance.

The second process to be described is the rate of *data flow*. Data can flow through the system only in accordance with the constraints imposed by the man and machine elements. Theoretically, a point can be reached where the rate of data input exceeds the capability of the system to act in response to it. The system may have various alternatives open to it under the condition of overload. One is a deterioration in performance; the system may actually become completely disorganized in the face of serious overload. Another alternative is the utilization of any storage facilities available to the system to delay the effects of the excessive input rates. Which of the alternatives is most appropriate has to be decided in terms of the least damaging compromise among the factors of deterioration in performance, allowable delay in processing time, and available storage capacity.

The model: cultural framework

As a background for the suggested development of a model in the cultural framework, we shall in this section enlarge upon three ideas: first, drawing out the implications of the fact that the machines in a system are cultural artifacts; second, considering the fact that the human comes to the system with some habits so strongly established that they resist all efforts at extinction; and third, considering the extension of some of the basic ideas to apply to groups of individuals in ways that the earlier ideas applied to single individuals.

With regard to our first point, it is of importance to characterize machines as products of human culture. It is not very meaningful to consider machines without also considering that they are made in accordance with cultural capabilities and are used in accordance with cultural patterns. A radar set is made in accordance with our store of engineering science and is used in the patterned activity of warfare, or air traffic control, or navigation. Little can be gained by the consideration of a radar set apart from its cultural determinants. Ultimately we must talk, not about the radar alone, but also about the scientific theory that governs its operation or the cultural patterns that govern its use.

The ultimate consequence is that systems study may become largely a social study. To coin a phrase, 'The proper study of systems is a study of culture.' In effect, this means that it is profitable to direct attention toward the impact of culture on the performance of man/machine systems. This assertion does not question the value or the validity of other approaches; it only highlights the present position.

While it is not the purpose of the present paper to elaborate on the concept of culture, it may be profitable to present some summary statements in order to avoid misunderstanding. Culture is defined to include the activities, the behavior patterns and the artifacts customary in the society. An example of these features of culture is found in the middle-class family at meals. This example is given with essentially the same comment in Gillen (1948). The behavior patterns are similar, the activity of dinner is a common one to an entire community, and the artifacts of knives, forks, spoons and dishes are remarkably similar from family to family. They are cultural features in that they are socially shared. In other words, the patterns of behavior, activities and artifacts must be meaningful to all the members of the society. Patterned behavior and activities are also present in the military situation. The custom of saluting, as well as the use of the knife and fork as customs, are meaningful to (shared by) the members of the society.

Customs are the elements of culture. Gillen defines custom as a habit 'which is socially learned, socially performed, and socially transmitted'. Artifacts are explainable in terms of the customs which produce them and they are understandable in terms of the customs employed in their use. It is characteristic of culture that it serves to facilitate human adaptation to the environment, both natural and social.

The second idea in our overall framework, the fact of strongly imbedded habits, is introduced to serve as a reminder that we may not wish to interfere with human habits that are hard to change. For instance, in the design of a communication system, we probably take advantage of the habit of speaking the English language in our culture rather than change this habit to one of speaking Esperanto. The choice is made on the basis of the consequent difficulties and complexities of any decision to change the strongly imbedded language habit.

There are, however, some peripheral language habits that we might choose to modify. The phonetic alphabet is a case in point. The phonetic alphabet is a set of English language words used to serve as redundant messages about letters of the alphabet. For instance, if the letter 'A' is to be transmitted, it is followed by 'Able'. It is thought that this redundance reduces the uncertainty about the original letter.

After the original 'Able, Baker . . .' code had been in use for some time, it was necessary to revise the code to eliminate confusion with certain foreign language words and pronunciations. Consequently, a revised phonetic code is now in use and is generally referred to as the new International Civil Aviation Organization or the 'Alpha, Bravo . . .' code. The efforts to establish this alphabet demonstrate an important principle: that it can be appropriate to intervene in an area of complex habit patterns, but the decision to intervene must be based on the choice between the relative habit pattern disruption and the estimated improvements in performance.

The final idea in our overall development is that of the application of our basic ideas to group interaction. Group is defined here as two or more individuals capable of interacting with each other in the psychological sense. Underlying group interaction is the fact that our society effectively provides patterned ways in which this interaction can take place. It is a characteristic function of culture to provide a kind of 'social efficiency' by establishing sets of appropriate behavior for various circumstances. In other words, the complex of culture and its elements of customs contribute to a certain amount of predictability, social harmony and order among members of society. It tends to eliminate wasteful behavior and in this sense provides 'social efficiency' since, in general, the individual is freed of the necessity for trial-and-error behavior by referring to the appropriate behavior as specified by the culture. Culture's role is, in part, to specify appropriate group interaction. There is an appropriate behavior of children towards adults that specifies the form of address for each towards the other, for instance, that allows each of them to predict in a general way how the other will, or at least should, interact. It is this specification of appropriate behavior that makes it possible for members of the society to meet on common terms with other members.

These group interactions fall into various constellations based on an arbitrary selection of the variables involved. One set of elementary constellations is based on the concept of *leadership*. If we define the group as having hierarchic patterns of interactions, we then may investigate the consequences of group behavior that arise from strong or weak leadership, or from the locus of authority in the formal group structure as compared with the informal structure. We may, moreover, manipulate the variables of leadership and authority in the experimental technique of examining the consequences of the controlled changes we make in the system.

We can distinguish at least three other concepts that serve as the core of behavior constellations. One of these is the *social dyad*, or the two-person interaction. Another is the *syndicate*, described as the group with *official authorization* to undertake some specific task. Experiments built around this concept would manipulate the variable of official authorization, its strength, its preciseness, or its explicitness, and examine the consequences on system performance. Finally, there is the concept of the *coalition*, described as a temporary, informal alliance for joint action. An experiment using this concept could, for example, provide structured situations that would permit coalitions to develop. The investigator's task might then be to generate principles for the choice of situations that lead to the most effective coalitions.

These group interactions of concern to us take place within the matrix of culturally patterned activities. We should use such knowledge of culture and its strongly entrenched customs as a set of guidelines for the study of systems and, further, to aid our understanding of men's performance within a system.

Now let us consider the numerous constraints that control the actions occurring in a system. We have enumerated some of them, namely, those imposed by human behavior and the rules under which social customs persist. At this point it is appropriate to categorize these constraints systematically to enlarge the utility of the system model we are developing. Three general types of control can be distinguished: the *coercion of nature*, the *policy of management*, and the *procedural customs*.

We may think of the natural environment (here called nature) as providing a set of limiting rules of system performance. As an obvious example, the law of gravitation is one constraint on system performance and the actions of the system are (with the exception of 'weightless' space flight) subject to the laws of motion and acceleration due to gravity. The attenuation of electromagnetic radiation, the physiological functions of the human organism, the availability of natural resources, and possibly even the weather, are all, in effect, controls by nature over system performance which it may be necessary to take into account in the function of a particular system.

Management policy similarly exerts control over system processes. This policy is generally established by the managers of the system in the form of rules that permit or prohibit certain actions. Obviously, the statement of policy by the system managers excludes a class of actions from the system's performance repertoire. This effectively controls the actions of the system in so far as it specifies that choices have to be made from some given universe of actions.

Finally, among the general categories of control are *procedural customs*. These are distinguished from statements of policy, in that policy attempts to say *what* things are appropriate to do, while customs determine *how* things are to be done. Procedural customs, moreover, can be thought of as operating rules. For example, the procedural customs in some air traffic control systems specify that traffic will be handled in turn by a pick-up controller, a feeder controller and a final approach controller. Once the procedures are established, the action structure of the system is controlled to the exclusion of alternate ways of operating the system. This control remains in effect unless the procedures themselves permit changes to be made under certain conditions or the system itself breaks down.

In addition to the general categories of controls, there are at least five other specific controls of interest (see Table 1). They are:

1. The authority structure.
2. The task requirements.
3. The value of the payoff.
4. The charter and its values.
5. The organization for communication.

The specific controls are discussed separately because they may be part of one or more than one general control category. For example, the authority structure may be imposed by a management policy or it may change from time to time in accordance with procedural custom. It is conceivable that the authority structure may be controlled in some cases by the coercion of nature, as in the case of the loss of the leader in battle.

Table 1 **Control antecedents and corresponding pattern consequences**

Control antecedents	Pattern consequences
Locus of authority and power hierarchy structure	Decision patterns
	Operating patterns
Task requirements	Alliance patterns
Value of payoff	Sanctions pattern
Charter and its values	Information flow
Communications organization	patterns

Ordinarily the elements in a system are faced with choice-making problems. The system, for example, may need to select one of a number of target signals. System performance may require that some action, A, be taken if the target is friendly and another action, B, be taken if the target is hostile. Who decides the answer to the critical question, 'Is the target friendly or hostile?' If we define authority as the power to make a choice, then there are at least three categorical authority structures for making the choice. The authority can be absolute and reside in a single person designated *absolute leader*. The authority can be delegated and reside in inferior members of the operating staff. The ultimate responsibility in this case, however, rests on the leader. Finally, the authority can be *distributed*, and reside in two or more members of the group. The typical example of completely distributed authority is a pure democracy. In the case of distributed authority, moreover, responsibility and authority are inseparable; that is, the persons making the decision are jointly and collectively responsible for its consequences.

Different authority structures lead to different decision patterns. Outcomes in groups controlled by an absolute leader are likely to be different from those in a group of distributed authority. Furthermore, the hierarchic structure of authority is an important element of control. It is of some consequence to know if the levels of authority are rigidly fixed, for in some cases only incidental decisions can be made by the inferior staff. The hierarchy may be so arranged that absolute authority resides at the top of the largest conceivable system organization as in a political dictatorship. It may be an important objective for the study of systems to evaluate the consequences of authority structures that locate authority at various levels in the system organization. We may wish to study the relative merits of the system that permits a fighter pilot to make absolute decisions in combat, versus the case where all decisions must be referred to the squadron commander.

We turn now to examine *task requirements* as a control of system performance. This control answers questions about who does what, when, and how. The task requirements are established, in large measure, by the procedural customs, and the operating patterns tend to come under the control of the kinds of task requirements the system contains. The dependence of the operating patterns on the task requirements has reference in turn to any other controls limiting the tasks themselves. In other words, the task boundaries may be set by management policy, by procedural customs, or by the coercion of nature. For example, the operating patterns in a GCI site grow out of the management policy that requires early warning about hostile aircraft. The task requirements are the practical mediators of this policy. The operating patterns are, in turn, the consequences of these requirements.

The third of the specific system controls is the *value of the payoff*. This value is composed of elements of evaluation about the outcomes of various courses of action. These evaluations serve as guides for compromises in behavior on the part of the system's staff. How, for example, may two fighter pilots carry out a combat mission? They may evaluate the consequences of protecting each other's flanks or not doing so. In the last analysis, the evaluation of the payoff for either course of action by the pilots may lead to some kind of alliance. To the extent that patterns of alliances grow out of this kind of evaluation of the quality of the payoff, we consider the value of the payoff a control and the alliance pattern its consequence.

The fourth of the specific system controls is the *charter and its values*. We define the charter as the authoritative statement of paramount values for the system, thus providing the standards for judging the value of the system performance. Defense of the nation, economical use of resources, destruction of the enemy, safety to friendly operating personnel, and desirability of heroism are typical matters that the charter makes statements about. The charter serves as a source of guidance for the applications of sanctions. The positive or motivating sanctions arise from the promises and the distribution of rewards for attaining the valued outcomes as stated in the charter. On the other hand, the negative or inhibiting sanctions arise from the threat and distribution of punishment for failing to attain the valued outcomes. Withholding of rewards may be punishing also.

Every member of the system staff is an agent for applying sanctions to any of the other members. Their understanding and acceptance of the statements of the charter, however, determine how strongly the sanctions will be applied. Intuitively, one feels that sanctions would be weak in a system where the charter is vague or where it is unpopular. It is to be remembered also that authority status enhances the strength of the sanctions, making the sanctions applied by the leader usually of more effect than those applied by other members of the staff. It is conceivable also that the strength of sanctions could increase through their uniform application by large proportions of the non-leadership staff, as in the case of revolt against authority.

The final specific system control we shall discuss is the *organization for communication*, that is, the way information is assembled, transmitted, and processed in the system for reducing equivocation. The communications organization is composed of elements that specify the communication links, the points between which the signals flow. The organization specifies also the quantity of information required by the system and available to the system. The organization may also specify the quality of the information required and available. All of these characteristics of the organization for communication exert an influence on the patterns of information flow.

In concluding this section, we need to draw out the implications of the particular way in which we have described the operation of a system. Of what profit is it to talk about the antecedent controls and the consequent patterns? For one thing, the descriptions are made in terms of definable variables. This way of characterizing the model is heuristic, and experimentation that may grow out of the questions suggested by the model may indeed supply profitable insights into actual systems operation, through appropriate manipulation of the antecedents and examination of the related consequences. It is this fundamental orientation that guided the development of the present system model. Through experimentation we seek the invariants that connect the pertinent antecedents with their consequences. The variables we have attempted to define and enumerate; the experimental work remains to be done.

Systems logic

The general procedure for systems study is to locate appropriate places in the system where we can intervene, and then to test the suitability of various ways of intervening. We can begin this general procedure by defining the output model, ω_D, in measurable terms and then by comparing ω_D with the system output, ω. The amount by which the output, ω, fails to coincide with this model, ω_D, is the system error, ω'. After intervening at some point in the system, we observe the changes in ω'.

A systematic technique is needed to simplify the choice of the method and of the place to intervene in the system. Are we interested, for example, in improving the data transformation process? If we intervene at this point, what kinds of changes may we make? The guidelines that we suggest are set forth in the following systems logic. This logic is a set of postulates and principles that may be used to guide our research.

This system logic focuses on the features of human behavior, in order that we may generate principles about the influence of human factors on systems performance. It goes without saying, of course, that there are other organizing schemes; the engineer, for example, may choose the servo loops as a point of focus. Because of the psychologist's concern with human behavior, we speak from this framework.

Postulate 1

Habits of behavior are formed or extinguished as a function of the kinds and immediacy of reinforcement or reward. Since we are not at this point intent on enumerating the precise conditions under which learning and extinction occur, it will be sufficient for our purposes to generalize as follows:

Habits are strengthened as a function of positive reinforcement or reward, or of the removal of negative reinforcement or a punishing condition. Presence of reward and punishment do not have exactly opposite effects on habit strength. A habit *may* be repressed or temporarily weakened by the judicious application of pain or punishment, but complete extinction takes place only when reward is withheld. It is also thought that the stronger the habit, the more resistant it will be to extinction if reward is finally withheld. The implications of Postulate 1 are set forth in the following principle:

Principle 1. A system should be arranged in order to provide rewards for appropriate behavior. Reinforcement, positive or negative, may be in the form of knowledge of results displayed in such a way that desirable habits are strengthened by the positive reward value of the information, and undesirable ones weakened by the removal of positive reward.

Postulate 2

In the cultural framework, the concept of *data mediation* takes on a more specific meaning. Transformation, we may recall, acts upon the data in an invariant way. If the linear transform results in the scaling of information dealing with the environment by a reduction factor of 1000 to 1, then this feature is fixed. Mediation, on the other hand, provides for interpretation. The human operator has some culturally determined ways of interpreting the *cue aspect* of the data he may be receiving. In short, it may be said that culturally derived habits determine *what* the man will do about the data. It is this *what-to-do* function we are concerned with in data mediation. The principle under which this function operates in the human operator is this: *mediation is determined by the operator's system of values, beliefs, and sanctions.*

For the purposes of our model we draw our definitions of values, beliefs and sanctions from Whiting and Child (1953) as follows: 'A belief is a custom whose response symbolizes some relationship between events.' A value is a 'custom whose response attributes goodness or badness to some event.' Positive and negative sanctions are, respectively, 'The customary rewarding and punishing behavior of others following the performance of a practice (or the failure to perform it when the occasion demands) . . .'

Behavior is guided in part by mediated cues. The organism reacts to a rather complex set of cues and stimulus patterns. The cues that are most similar to those present in previous situations tend to have the strongest effect on behavior. Likewise, the cues that are compatible with the behavior patterns of the organism are likely to have strong effects on behavior. Cues that contradict our beliefs are frequently weak in their influence on behavior. In other words, a cue has to be credible before we commit ourselves to act in terms of it. In addition to these influences, we can also be taught to notice some class of distinguishable cues and we are said to have 'learned to pay attention'. It is thought also that habit evocation is irregular in the presence of vague or equivocal cues. Postulate 2 supports the following principle:

Principle 2. Cues provided by the system should be structured in such a way as to be compatible with existing values, beliefs and sanctions; moreover, the cues themselves should be unequivocal.

Postulate 3

One of the functions of the machine elements of a system is to couple the man to his surroundings. In general, this coupling depends on symbolic transformations of the data and of the system output. For example, data about speed are transformed into the symbolic language of numbers. Directions for manipulations, for instance, symbolized by arrows, are transformed by machine linkages into movement of wheels or flaps. In either case, the symbols are meaningful because of the culture in which we live.

The culture, in fact, determines our responses to these symbols. To use an extreme example, Egyptian hieroglyphics are inappropriate as symbols in an English language culture unless we assign to each of them an arbitrary social meaning or train people in their native Egyptian meanings.

Principle 3. Symbolic transformations should be consistent with the cultural biases of the human operator.

Postulate 4

Learning of desirable habits is enhanced by *warnings, assistance in generalization, and the provision of models for imitation.* Examples from experience certify the usefulness of this postulate. Prior warnings assist in directing attention to appropriate cues; insights come easier if someone points out the generalizations, and it is well known that we may learn a desirable golf swing if the pro guides our arms for a while.

Principle 4. Systems should enhance learning by providing appropriate warnings, assistance in generalization, and models for imitation.

Postulate 5

Strongest reinforcement comes from immediate rewards. It is known that the value of reinforcement is lessened as a function of time between response and reward. The attenuation is complete if the delay is infinitely long. In other words, infinite delay of reward is equivalent to providing no rewards.

Principle 5. Knowledge of results must follow response with minimum delay in time.

Postulate 6

The rate at which the organism can process data is limited. In general, rates of input greater than the processing rate have to be stored or wasted.

Principle 6. If the system input is such as to overload the operator, storage facilities should be provided for the excess flow of input data that cannot be handled by the operator.

References

GILLEN, J. (1948), *Ways of Men*, Appleton-Century-Crofts.
WHITING, J. M., and CHILD, I. L. (1953), *Child Training and Personality: A Cross-Cultural Study*, Yale University Press.

6 Arthur D. Hall

Some Fundamental Concepts of Systems Engineering

Excerpt from A. D. Hall, *A Methodology for Systems Engineering*, Van Nostrand, 1962, pp. 59–74.

Orientation

The most useful method of defining systems engineering is to present a basic and detailed description of the process. Now it is time to begin the description of the process free from encumbrance of the technical details of a case history. This chapter will define and discuss a few simple but fundamental concepts and propositions which lie at the foundation of systems engineering.

In preview of this reading, the primary concept to be discussed is that of *system*. Closely related secondary concepts to be discussed are: *environment, need, planning* and *creative process*. It happens that certain properties apply to systems in general, irrespective of the nature of the systems or of the fields in which they are normally studied. While it is true that not all of the most general properties are useful in an operational sense for applied work, they have considerable conceptual value in more deeply understanding creative or developmental processes. This fact is the real justification for including them in this chapter.

The author emphatically does not believe that there is a neat set of rules which permit their skilful manipulator to pass from a vaguely felt need for a new system to the finished system in a perfect step-by-step sequence. The problems of defining the need, creative synthesis, wise value judgments, and of other crucial operations have never been solved by rules alone. Yet there are certain definite steps in the systems engineering process, a few significant principles and viewpoints, and a few more-or-less efficient approaches to certain recurring operations in the process. Furthermore, there is good evidence that, while knowledge of these matters may not permit a poorly endowed man to do a superlative job, such knowledge will enable a good man to do a better job in less time.

Definition of system

Unfortunately, the word *system* has many colloquial meanings, some of which have no place in a scientific discussion. To exclude such meanings, we state the following definition:

A system is a set of objects with relationships between the objects and between their attributes.

Objects are simply the parts or components of a system, and these parts are unlimited in variety. Systems may consist of atoms, stars, switches, springs, wires, bones, neurons, genes, gases, mathematical variables, equations, laws and processes.

Attributes are properties of objects. For example, in the preceding cases the objects listed have (among others) the following attributes:

1. Stars: temperature, distances from other stars.
2. Switches: speed of operation, state.
3. Springs: spring tension, displacement.
4. Wires: tensile strength, electrical resistance.

Relationships tie the system together. In fact, the many kinds of relationships (causal, logical, random etc.) make the notion of 'system' useful.

For any given set of objects it is impossible to say that no interrelationships exist since, for example, for a particular physical system, one could always consider as relationships the distances between pairs of the objects. The relationships to be considered in the context of a given set of objects depend on the problem at hand, important or interesting relationships being included, trivial or unessential relationships excluded. The decision as to which relationships are important and which trivial is up to the person dealing with the problem; i.e. the question of triviality is relative to one's interest. Although a systems engineer may wish to restrict the term *system* to 'really complex assemblies', it is just as valid for a chemist to refer to the structure of a nucleic acid molecule as a system.

Definition of environment

Environment for our purposes can be best defined, in a manner quite similar to that used to define system, as follows:

For a given system, the environment is the set of all objects outside the system: (1) a change in whose attributes affect the system and (2) whose attributes are changed by the behavior of the system.

The statement above invites the natural question of when an object belongs to a system and when it belongs to the environment; for if an object reacts with a system in the way described above, should it not be considered a part of the system? The answer is by no means definite. In a sense, a system together with its environment makes up the universe of all things of interest in a given context. Subdivision of this universe into two sets, system and environment, can be done in many ways which are quite arbitrary. Ultimately it depends on the intentions of the one who is studying the particular universe as to which of the possible configurations of objects is to be taken as the system. The reader may readily supply examples to illustrate this idea.

Systems and their environments

The general problem of specifying the environment of a given system is far from trivial. To specify an environment completely, one needs to know all the factors that affect or are affected by a system; this problem is in general as difficult as the complete specification of the system itself. As in any scientific activity, one includes in the universe of system and environment all those objects which he feels are the most important, describes the interrelationships as thoroughly as possible, and pays closest attention to those attributes of most interest, neglecting those attributes which do not play essential roles. This method of *abstraction* or *idealization* works rather well in physics and chemistry; massless springs, frictionless air, perfect gases etc. are commonplace assumptions and simplify greatly the description and analysis of mechanical and thermodynamical universes. Biologists, sociologists, economists, psychologists and other scientists interested in animate systems and their behavior are not so fortunate. In these fields it is no mean task to pick out the essential variables from the nonessential; that is, specification of the universe and subsequent dichotomization into system and environment is in itself, apart from analysis of the interrelationships, a problem of fundamental complexity.

For the purposes of this chapter we will be interested in those environmental factors which lead to requirements on man-made systems. These factors will differ somewhat according to the specific field of design and the kind of organization engaged in systems engineering. This is why it is essential for each organization to isolate and study those factors peculiar to it. Some factors which appear universal are:

1. The state of technology.
2. The natural environment (climate, plant life, etc.).
3. Organization policies.
4. Economic conditions for new systems.
5. Human factors.

It is important to see that physical systems do not exist merely *in* an environment; they exist *by means of* an environment. This is easy to see for living things because life involves energy exchange between the organism and its environment. For example, the difference between existing in and by means of an environment is not just that a fish lives *in* the water, but that the characteristic functions of the fish are conditioned by the special way in which water enters into its activities. In the creation of a data-processing system to satisfy some set of needs, we do not only design the system to put *in* the environment, rather we find the system is *determined by* the environment. In fact the success of the design is measured by the *closeness of fit*, i.e. the degree of integration with the environment. There are many practical cases where the fit with respect to a particular environmental factor is deliberately compromised to get the optimum fit with the total environment. These thoughts suggest the main reason for studying the environment – the better the environment is understood and evaluated, the more closely can the system be engineered.

From the foregoing we perceive one of the central objectives of the entire developmental sequence, which is to locate optimally two functional boundaries, or interfaces:

1. The boundary setting off the universe of things of interest in a given problem.
2. The boundary between the system and the environment.

Subsystems

It is clear from the definition of system and environment that any given system can be further divided into subsystems. Objects belonging to one subsystem may well be considered as parts of the environment of another subsystem. Consideration of a subsystem, of course, entails a new set of relationships. The behavior of the subsystem might not be completely analogous with that of the original system. Von Bertalanffy (1950a) refers to the property of *hierarchical order* of systems: this is simply the idea expressed above regarding the partition of systems into subsystems. Alternatively, we may say that the elements of a system may themselves be systems of lower order.

Hierarchies of physical systems, plans and goals

The concept of hierarchical order has many useful applications. The following are three important kinds of systems: a physical system, a system of plans and a system of goals.

The largest single physical communications system in the world, the United States–Canadian telephone network, exhibits hierarchical order as one of its most obvious properties. The hierarchy consists of five ranks, obtained by dividing the geographical area five times. The lowest ranking area is served by a Class 5 end office, which is an ordinary local telephone exchange at which the customer's lines terminate. The long-distance telephone calls from several of these end offices are processed by a toll center which has rank 4 in the system, and which passes traffic to the next higher ranking office called a primary center. Most of the traffic is passed from primary center to primary center, but in case there is no available channel the call is steered to the next higher ranking office in the distant area, called a sectional center. Depending on the amount of traffic in the system (or possibly other conditions), the call may 'climb' the hierarchy to the highest rank in the originating area called the regional center. The organization is given by Figure 1. Each rank has different functions, and several kinds of complex switching equipment are used to implement the functions, but a unifying property of the whole is hierarchical order.

An important observation on planning derives from the concept of hierarchical order; it is that plans may occur in hierarchies, too. In systems engineering a general objective is implemented by a whole series of successively more detailed plans. One of the functions of the systems engineer is to ensure the internal consistency and integration of plans at different hierarchical levels.

It is intriguing and useful to note that since goals are implied by plans, they, too, occur in hierarchies. This fact leads to an important step in setting objectives for systems engineering and development projects; it is to test for consistency of a set of objectives at one level with those at a higher level.

Some macroscopic properties of systems

So far we have been talking about systems as if by implication there were in the background some sort of unified theory of systems. Actually, there is as yet no such theory, although attempts have been made at one, for example, by the 'general systems' theorists. It is always a good idea when considering such

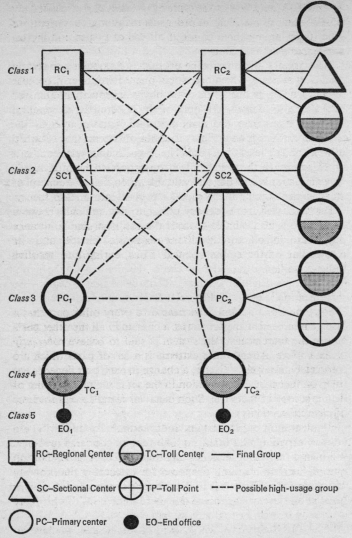

Class 1   RC₁   RC₂

Class 2   SC1   SC2

Class 3   PC₁   PC₂

Class 4   TC₁   TC₂

Class 5   EO₁   EO₂

RC–Regional Center   TC–Toll Center   —— Final Group

SC–Sectional Center   TP–Toll Point   --- Possible high-usage group

PC–Primary center   EO–End office

Figure 1 Plan for routing telephone traffic in the direct distance dialing system

general theories to be sure the types of system under discussion are clearly understood and, where generalizations to systems of other types are claimed, to see if all the analogies and correspondences used are valid.

Nevertheless, there are some properties that belong to certain classes of systems, which are worth mentioning briefly, because sometimes very difficult system problems are greatly illuminated by looking at them in the light of the appropriate generalized property. Also, there are some useful analogies concerning the behavior and properties of certain types of systems that often aid in analysis, at least conceptually, of particular systems. As a notable example, the concept of entropy, useful in thermodynamic systems, has an interesting and valuable analogue in the concept of entropy as defined for message sources in information theory. Other familiar examples are found in the close analogies between electrical, mechanical and acoustical systems, a simple instance being an inductance-capacitance-resistance circuit and its mechanical analogue, the coupled mass, spring and resistive dashpot respectively.

Wholeness and independence

If every part of a system is so related to every other part that a change in a particular part causes a change in all the other parts and in the total system, the system is said to behave *coherently* or as a *whole*. At the other extreme is a set of parts which are completely unrelated – that is, a change in each part depends entirely on that part. The variation in the set is the physical sum of the variations of the parts. Such behavior is called *independence* or *physical additivity*.

Wholeness (or coherence) and independence (or additivity) are not two separate properties, but extremes of the same property. Wholeness and independence are matters of degree, but no sensible method has been proposed for measuring the property on a ratio scale. Nevertheless, the property is conceptually useful. Because all systems have some degree of wholeness, this property is used by one writer to define system.

Since all systems have wholeness in some degree, we have no difficulty illustrating the property. Such systems as passive electrical networks and their mechanical analogues are very coherent. At the other extreme we have difficulty finding examples. The term 'heap' or 'complex' is sometimes used to describe a set of parts which are mutually independent, and the term 'system' is used only when some degree of wholeness exists. We prefer to call sets of parts with complete independence *factorable systems* because, as noted before, it is impossible to deny systematic relationships in odds and ends or even in a heap of sand.

Progressive factorization

The concepts of wholeness and additivity can be used to define another property often observed in physical systems. Most physical systems change with time. If these changes lead to a gradual transition from wholeness to independence, the system is said to undergo *progressive factorization*. For illustration, consider the two real variables x_1 and x_2 satisfying the system of two linear equations:

$$a_1x_1 + a_2x_2 = c_1$$
$$b_1x_1 + b_2x_2 = c_2$$

Now let the 'mutual' or 'transfer' terms a_2 and b_1 become functions of time. If these terms decrease to zero as a limit, we will have two independent systems represented by the equations, or we can say that the larger system, consisting of two simultaneous equations, becomes 'factorable.' Systems in which the parameters are functions of time are variously called time-varying, nonstate-determined, absolute, and (particularly in reference to stochastic systems) nonstationary.

We can distinguish two kinds of progressive factorization. The first and simplest kind, illustrated above, corresponds to decay. It is as though, through much handling, the parts of a jig-saw puzzle become so rounded that a given piece no longer fits the other pieces better than another. Or suppose an automobile were deprived of maintenance. The engine would wear out, tires would rot, and eventually the parts will no longer behave as a system.

The second kind of progressive factorization corresponds to growth. The system changes in the direction of increasing division into subsystems and subsubsystems, or differentiation of functions. This kind of factorization appears in systems involving some creative process or in evolutionary and developmental processes. An example is embryonic development, in which the germ passes from wholeness to a state where it behaves like a sum of regions which develop independently into specialized organs. Another example occurs during the creation of a new man-made system, when an idea appears (or a need is defined) and the original conception of a system is factored through planning effort into subsystems whose design and development eventually proceed almost independently.

Progressive systematization

This is simply the opposite of progressive factorization, in which there is change toward wholeness. It may consist of strengthening of pre-existing relations among the parts, the development of relations among parts previously unrelated, the gradual addition of parts and relations to a system, or some combination of these changes. As an example, consider the development of the long-distance telephone network. First, local telephone exchanges sprang up about the country. Then exchanges were joined with trunk lines. As transmission techniques improved, more exchanges were added at greater distances. Later, distance dialing was added, placing the network at the command of operators and eventually at the command of customers. The record has been one of increasing unification of the whole system.

It is possible for progressive factorization and systematization to occur in the same system. These two processes can occur simultaneously and go on indefinitely so that the system can exist in some kind of steady state as with the processes of anabolism and catabolism in the human body. These processes can also occur sequentially. Consider the early history of America during which groups of people colonized various parts of the country. These groups became more and more independent of their parent countries. Gradually, the new country became more coherent as further interchanges occurred between the groups, common traditions were established, and a new government formed.

Centralization

A *centralized system* is one in which one subsystem plays a major or dominant role in the operation of the system. We may call this the *leading part* or say that the system is *centered* around this part. A small change in the leading part will then be reflected throughout the system, causing considerable change. It is like a trigger with a small change being amplified in the total system. An example from politics might be a totalitarian regime, the decisions of an autocrat affecting behavior of the entire system.

Either progressive factorization or progressive systematization may be accompanied by *progressive centralization*; as the system evolves, one part emerges as a central and controlling agency.

In the case of embryonic development previously noted, factorization does not proceed to the limit for several reasons, the most important perhaps being that the brain emerges as the controlling and unifying part. The same tendency appears in the evolution of man-made systems, such as the stational telephone network and national defense systems in which more and more subsystems are designed as 'slaves' and placed under control of 'brains' or computers.

The concept of the centralized system yields the important principle that the more a system is centralized, the more the leading part must be protected against damage from unstable environmental factors. Thus a dictator requires an elaborate security system to keep himself in power. So also in the human body is the brain maintained in a stable environment by the most elaborate network of feedback loops for holding constant the temperature, sugar, water, oxygen, pH and similar factors.

The concepts of centralization and hierarchical order are related. A completely decentralized system has only one level of organization, each part of the system sharing responsibility for carrying out the set of functions. Higher degrees of centralization accompany more levels of organization. Just as in social systems such as government and business organizations, many design questions are associated with the degree of centralization that a system should have to perform optimally. Examples: how many levels of organization should the system have? How many subsystems should report to a given level? Which functions should be carried out at a given level, and in what part of the system should a given function be located? What kinds of links are needed for the flow of information or materials to tie together the various levels of the system? All of these questions were asked and answered in the design of the telephone distance dialing system noted above. Such questions represent a recurring theme in planning and design and they bear repeating at each level as a guide to the design of all lower levels.

The foregoing system properties are summarized by Figure 2. The two axes are purposely left unscaled to emphasize the point that in our present state of knowledge the two concepts of wholeness and centralization are not measurable on ratio scales.[1]

1. This statement may be unnecessarily cautious. It is clear that each scale has a natural zero. It may well be that the concepts of connectivity index and centrality index as used in graph theory, when properly modified to account for the strengths of the connections among the parts, may be useful measures of wholeness and centralization respectively.

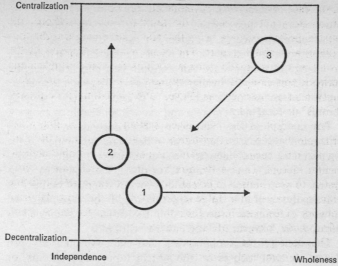

Figure 2 Macroscopic properties of systems

Nevertheless, the graph shows three systems each undergoing a different kind of change. System 1 shows progressive systematization while the degree of centralization remains constant. System 2 shows progressive centralization while the degree of wholeness remains constant. System 3 shows simultaneous progressive factorization and decentralization.

Natural and man-made systems

To enhance the meaning of system we may distinguish natural systems and man-made systems. Systems engineers are directly interested in man-made systems; however, the environment of man-made systems often contains natural systems which also require investigation, since their properties interact with the system under study. Furthermore, there are certain properties that both types of systems have in common; man-made systems are often copies of natural systems or at least are constructed to perform analogous functions.

Natural systems

The description of these is the task of the astronomer, physicist, chemist, biologist, physiologist or sociologist, and again what one can say about a given natural system depends on the number of essential variables involved.

Arthur D. Hall 113

Open and closed systems

Most organic systems are *open*, meaning they exchange energy with their environments. A system is *closed* if there is no import or export of information, heat or physical materials, and therefore no change of components. An example is a chemical reaction taking place in a sealed insulated container; this example suggests that one of the uses of the concept of a closed system is to simplify the model of a physical system and make it amenable to analysis.

Whether a given system is open or closed depends on how much of the universe is included in the system and how much in the environment. By adjoining to the system that part of the environment with which an exchange takes place, the system becomes closed. For instance, the second law of thermodynamics is universally applicable to closed systems; it only *seems* to be violated for organic processes. For the organic system *and its environment*, however, the second law still holds.

In studying either a natural or a man-made system as an open system, interest focuses on the system inputs and outputs, or *throughputs* as they may be called, since it is the transport of energy through the system which allows us to get work from it.

Consider a digital computer as an open system. This system will exist in steady-state if we supply 'high' forms of energy such as electric power, human energy for maintenance, spare parts etc. and if we remove 'lower' forms of energy such as heat or used-up tubes. In this condition the computer can do work by accepting information at its input in the form of data and a program of directions, transforming the data as specified by the program, and presenting the results at its output.

Powerful aid comes in analyzing open systems by recognizing their *network* aspects. Transportation, communication, pipeline and power distribution systems all have obvious network characteristics. In the human body the nervous and blood distribution systems are equally obvious networks. A little thought shows that most open systems have network characteristics.

Considerable mathematical theory of networks is available. Analysis can be applied to the nodes (which can embody complex equipment), to branches (which may be complex systems like a radio relay) or to the network as a whole, to deduce the outputs for a given input for example. The quantities studied may include the rates of flow of the throughputs, their concentrations, their time and frequency distributions, and their kinds. Some of the tools that have been applied to network problems include graph theory (a branch of topology), signal-flow graph theory (a branch of graph theory), traffic (or queueing) theory and linear programming.

Adaptive systems

Many natural systems, especially living ones, show a quality usually called *adaptation*. That is, they possess the ability to react to their environments in a way that is favorable, in some sense, to the continued operation of the systems. It is as though systems of this type have some prearranged 'end', and the behavior of the system is such that it is led to this end despite unfavorable environmental conditions. The 'end' might be mere survival; evolutionary theory is based heavily on the notion of adaptation to environment. Since adaptation may permanently (irreversibly) change the operation or use of the system, such systems are also called *systems with variable utilization*.

There are many examples of adaptive behavior in the body. Many of these are mechanisms that tend to keep within certain physiological limits various bodily conditions such as temperature and physical balance. Mechanisms of this sort are sometimes called homeostatic mechanisms. One example, called thermostasis, is the inborn reaction to cold by shivering, or the tendency to resist a drop in body temperature by a compensating movement producing warmth. Closely related to the concepts of adaptation, learning and evolution is the notion of stability.

Stable systems

A system is stable with respect to certain of its variables if these variables tend to remain within defined limits. The man-made thermostat is an example of a device to ensure stability in the temperature of a heating system; the notion of stability is familiar also in mechanics and in the communications field. An adaptive system maintains stability for all those variables which must, for favorable operation, remain within limits. In physiology, motor coordination is intimately connected with stability; clumsiness, tremor and ataxia are examples of deficient or impaired motor coordination and instability.

Systems with feedback

Certain systems have the property to reintroduce a portion of their outputs or behavior at their inputs to affect succeeding outputs. Such systems are familiar enough to the communications engineer; servomechanisms in general are man-made systems utilizing the principle of feedback. Systems with feedback occur quite frequently in nature as well; posture control in the human body is an example. It is a well-known fact that the nature, polarity, degree and time lag of feedback in a system have a decisive effect on the stability or instability of the system.

Very complex systems may have several interacting feedback loops. When a system exceeds the limits of control of one feedback loop, if it switches to another loop with different limits the system is said to show *ultrastability*, a term due to Ashby (1952). The TD-2 system provides an example: when a channel fades below the range of the AGC system, a new channel operating at a different frequency is switched in; when the fade is over, the original channel is switched back in.

Man-made systems

Man-made systems exhibit many of the properties possessed by natural systems; simple notions such as wholeness, factorization and additivity have meaning for both types of system. On the other hand, it has been only recently that man-made machines have shown what might be termed adaptive behavior even on a modest scale. Other kinds of man-made systems, such as language and systems of social organization, have always shown adaptive behavior.

Adaptation for man-made systems is not strictly analogous to that for natural systems; what might be considered mystical behavior on the part of a natural system is perfectly explainable for the man-made system. Any seemingly purposeful or intelligent behavior on the part of a machine has been built into it by its designer. Also, adaptive behavior on the part of a machine is not to ensure the survival of the machine necessarily, but instead to ensure a specified performance in some respect.

Perhaps the concepts of systems with feedback, ultrastability and adaptation may be clarified by reference to block diagrams of some man-made systems. Figure 3 is the basic feedback configuration. The process to be controlled has a completely determined relationship, or transfer function, between its input (the actuating signal) and its output. The transducer is simply a measuring device used to measure the output and generate a feedback signal which is subtracted from the input to produce an error signal. The controller changes the error signal into a

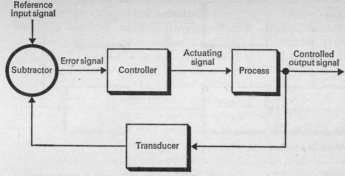

Figure 3 A basic feedback system

form needed to drive the process; it may be as simple as an amplifier.

Figure 4 is just one of several types of ultrastable system. Two new functions have been added to the basic feedback system. First, a threshold detector monitors the input, and when it finds that the input signal exceeds the limits for which the feedback loop using transducer number one was designed, it operates the two switches which disconnect transducer number one and connect transducer number two. The switches are restored to normal when the input returns to its normal range.

Figure 5 is the basic adaptive system. In place of a simple threshold detector is a more versatile function to analyse the input signal in the way needed; it may average the input, measure its derivative, or perform a similar operation. In place of

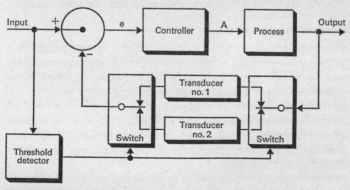

Figure 4 A type of ultrastable system

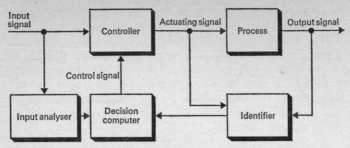

Figure 5 The basic adaptive system

the transducer is an identifier fed by *both* the output and activating signals; its role is to measure continuously the transfer function of the process. The resulting information together with the analyzed input signal is fed into a decision computer, which then decides upon the best activating signal for the controller to produce. In other words, the *design of the system automatically adjusts* as the process characteristics change due to changes in one or more environmental factors. This is in sharp contrast to the basic feedback system, whose *design is fixed* once and for all by the designer who had studied the process to identify its characteristics. From this we see that the ultrastable system is, at least in principle, only a special case of the adaptive system, since the decision computer could be programmed to produce a step function of the type used in the ultrastable system.

There are, in addition to the differences above, some further aspects of man-made systems that seem to have less bearing on natural systems. Among these are compatibility and optimization.

Compatibility (harmony)

Often the problem arises of constructing a system to match a given environment, or (what amounts to virtually the same thing) of adding new parts to already existing systems. There is no guarantee that a system constructed for a given purpose will function properly if its environment is changed (e.g. not all fountain pens write under water). Similarly, two systems independently might be quite satisfactory in certain respects, but in tandem could have completely different and not necessarily favorable characteristics.

Systems may be compared as to the degree of compatibility. In terms of a high-fidelity system, we might consider as an example the problem of matching a speaker to the rest of the outfit. Different speakers would function with varying degrees of success; some of the environmental factors might be the size of the room or the amount of money available to spend on the speaker. A speaker with perfectly matched impedance and excellent mechanical construction might produce beautiful results in the given setting, but if it cost a few thousand dollars it could easily be called incompatible with respect to at least one environmental factor.

Optimization

Compatibility considerations lead naturally to the problem of optimization. As the term implies, *optimization* means securing the best fit between the system and its environment. There are several other useful ways to put this. We might say that an open system is optimum if the system transforms its inputs into the most valuable outputs. There are some advantages of referring to desired outputs as objectives of the system; one may then speak of optimizing a system with respect to its objectives. Contrary to the way the term is often used, the operation of optimizing a system involves more than one function. Instead, it involves the solution of a complete problem including problem definition, choosing objectives, system synthesis, system analysis and selection of the best system.

Optimizing a subsystem with respect to its objectives, or optimizing the total system with respect to a subset of objectives is called *sub-optimization*. Separate optimization of all subsystems of a system does not guarantee an optimum system, because of interactions. Interactions may also prevent separate optimizations with respect to all subsets of (less than all) objectives from yielding a total optimum. For example, if the objective of a particular system design is minimum cost, the separate achievement of minimum manufacturing and installation costs will not ensure the prime objective, unless these are the only objectives and are independent.

Systems with randomness

In either natural or man-made systems it is sometimes necessary to take random behavior into account. There is a variance of opinion as to what randomness means and when to introduce it in the analysis of a system. In practice it is usually introduced as a factor when the variables that may affect a given attribute are so great in number or so inaccessible that there is no choice but to consider behavior as subject to chance. One example is the noise in an electron tube due to random emission of electrons from the cathode.

Random variables enter at both microscopic and macroscopic levels. Statistical mechanics and modern physics are both dependent on assumptions of microscopic randomness. Economic conditions and numbers of potential customers are macroscopic factors also subject to chance fluctuation. The operation of some systems with randomness can best be described in terms of stochastic processes (also called random processes or time series).

References

ASHBY, W. R. (1952), *Design for a Brain*, Chapman & Hall.
BERTALANFFY, L. VON (1950a), 'An outline of general systems theory', *J. Phil. Sci.*, no. 1.
BERTALANFFY, L. VON (1950b), 'The theory of open systems in physics and biology', *Science*, no. 3.
BERTALANFFY, L. VON, and RAPAPORT, A. (eds.) (1957), ' General Systems ' *Yearbook of the Society for the Advancement of General Systems Theory*, Braun & Brumfield.
NEWMAN, W. H. (1951), *Administrative Action*, Prentice-Hall.
STEBBING, L. S. (1930), *A Modern Introduction to Logic*, Cromwell.

7 E. S. Quade

Military Systems Analysis

E. S. Quade, 'Military systems analysis', The RAND Corporation, 1963, pp. 1–29.

The analysis of weapons and strategies for future wars presents a new kind of problem, different in a practical sense from any treated by operations analysts in the Second World War, or even in the Korean War. The aim in planning may now be how to deter war or even how to disarm with security, as well as how to wage war. A broad context, a rapid rate of technological change and a resourceful enemy clothed in secrecy make extremely hazardous any prediction of the environment – usually five, ten or more years in the future – in which the weapons and strategies are to be used, and the effect of their introduction into that environment. In this area of long-range military planning, as opposed to the operational use of given military units or weapons, piecemeal component optimizations and cost-effectiveness comparisons of competing postures and strategies must be replaced by an over-all treatment in which emphasis is placed on an integrated simultaneous consideration of all the major relevant factors.

Systems analysis, that is, analysis to suggest a course of action by systematically examining the costs, effectiveness and risks of alternative policies or strategies – and *designing additional ones if those examined are found wanting* – represents an approach to, or way of looking at, complex problems of choice under uncertainty. It was developed originally to deal with long-range military problems but is now used extensively by managers and engineers of large industrial enterprises, such as telephone companies and producers and distributors of electric power. It offers a means of discovering how to design or to make effective use of a technologically complex structure in which the different components may have apparently conflicting objectives; that is, an approach to finding the best balance among risks, objectives and cost. Its purpose is to place each element in its proper context so that in the end the system as a whole may attain its objectives with a minimal expenditure of resources.

It was not the systematic approach but the subject matter which originally suggested the name. The first military studies after the Second World War were primarily concerned with weapon systems. Evaluations undertaken to enable a decision maker to choose among systems, to discover whether a given system could accomplish the desired objectives, or to set up a framework within which tests of the system could be prepared were naturally called 'systems analysis'. With slightly different emphasis, the terms 'systems research', 'systems design', 'systems engineering' and, lately, 'operations research' are also used.

As an example of a relatively narrow problem in which a systems approach might be helpful, let us examine one which might arise in choosing a next-generation air defense missile from among several possible configurations. Consider, for example, guidance and control. Without taking a 'system' point of view, it might seem obvious that if the accuracy of our missile can be improved, the result will be more enemy missiles or planes shot down. It does not follow at all, however, that the missile with the highest accuracy will necessarily maximize the effectiveness of the over-all defense system or, for that matter, will even give the missile component of the system its highest potential for killing enemy vehicles. Any numerical values which measure the kill capability of a missile-defense system must depend on at least four factors: first, the number of missile emplacements within whose range the invaders must fly; second, the number of missiles that can be launched during the time the enemy is within range; third, the probability that a given missile will be operative; and fourth, the probability that an operative missile kills its target. An increase in the accuracy of the missile would probably increase this fourth factor. But this would result in an overall increase of kills only if the values of the other factors were not materially lowered by whatever change was necessary to bring about the increase in accuracy.

If, for example, additional guidance and control equipment were added to the missile to improve its accuracy, the resulting increase in weight might reduce the range or the speed of the missile. This in turn could reduce the number of missiles that might be launched in an engagement. Also, the greater complexity of more accurate guidance equipment might degrade the reliability. Consequently, in spite of the increased accuracy, the overall effectiveness might be reduced. Moreover, there is the very likely possibility that the more accurate missile might cost more and, since the total expenditure is certainly a constraint, the purchase of missiles which are individually more accurate might lead to fewer launching sites and fewer missiles.

Indeed, in these days of deterrent weapons, certain less obvious factors – for example, the state of readiness, the vulnerability and the susceptibility to countermeasures – may contribute equally to deterrence and be the items which dominate the costs of the system. The certainty with which a weapon can be fired after an attack may be more essential than its accuracy. Operational and logistic factors such as mobility, data requirements, communications, supplies, maintenance, personnel and training must all be considered in a systems approach. For example, before deciding to use an unusual substance as a fuel, on the grounds that it would enhance the range of the missile, the logistical implications in the decision must be investigated. The fuel may be so toxic that it will require inordinately complex handling for supply, transport and storage. If so, the overall performance of the system may be degraded, or the costs raised, in spite of any increased range or speed that might develop.

Thus, in this problem of choosing an air-defense missile, a systems approach is indicated. The context must be broad enough to embrace everything pertinent to all the alternative systems. The analysis would ordinarily take one or the other of two equivalent forms. For a given desired level of military effectiveness, the systems analyst might attempt to determine which alternative, or combination of alternatives, will imply the least cost. Or, for a specified budget level, he might try to find out which alternative, or combination of alternatives, will maximize effectiveness. In either case, the total systems analysis would require numerous substudies – for example, operations research to investigate problems of deployment or logistics, cost analysis to estimate the dollar costs of the several alternatives, and possibly even war gaming to suggest enemy penetration tactics.

The simplest category of systems analysis involves a choice from within a class of essentially similar alternatives. The problem of choosing the next-generation air-defense missile belongs to this category. The possible alternative missile systems may differ widely with respect to accuracy, range, payload and certain other characteristics, such as alert status. But they are likely to be similar in certain fundamental aspects in which the uncertainties are the greatest – for example, in the estimates associated with how far their performance in combat will fall below that of the proving ground, in enemy reactions to their development and use, and in their logistics and support problems. Since they are essentially similar means for accomplishing the same objectives, and are associated with the same time period, many uncertainties are likely to affect all designs in the same direction and approximately to the same extent. In analysis in which the alternatives are relatively similar, it is easier to take uncertainty into account and to apply measures to alleviate its consequences; also, one feels that the failure to handle this factor adequately is not so likely to invalidate the analysis.

On the other hand, a broader problem for systems analysis might involve the design of an entire air-defense system to protect the United States from damage. This would be difficult, but not merely because of the wider context involved. The value of an air-defense system is measured by more than its ability to prevent damage in event of a surprise attack which begins all-out war; for example, in peacetime it polices our borders and prevents intrusions. Better protection, however, results from preventing war, or, if war comes, from keeping it away from our country. Doing this depends on offensive power and national policy for using it – and on air defense. Thus the problem of finding agreement on a working basis for objectives and criteria is not likely to be an easy one.

Even after criteria and objectives have been tentatively set, considerable practical difficulties remain. Here, alternative sub-systems with complementary but essentially different tasks, such as radar and anti-missile missiles, would compete for resources. Moreover, even with weapons which have essentially the same objectives, say air-defense missiles for point defense and those designed for area coverage, new difficulties arise because such factors as the warning times required for their employment, their differing utilities under different enemy tactics, and even the support structure may be entirely different. The level of knowledge about the various ingredients will be different. Of course, for weapons such as aircraft there is much past experience to guide the investigator with respect to such things as maintenance requirements, reliability, and the like. For missiles, this backlog of experience does not exist. But even more serious are the effects of uncertainties about alternatives which contribute to damage reduction in entirely different ways – say, shelters and alert missiles. Further, since (as we mentioned earlier) a better way to prevent damage is not to have a war, the analysis has to consider also how these elements affect the likelihood of war, as well as the chances of survival if war comes.

It is not easy to tell someone how to carry out systems analysis. We lack adequate theory to guide us. The attention of the practitioners, when it has turned to methods, has been focused mainly on the development of mathematical techniques. This attention has met with great success. Models have now become easier to manipulate, even with many more variables represented. Computational obstacles cause comparatively little difficulty. It is the philosophical problems, such as occur in providing assurance that the model is meaningful, in devising schemes to compensate for uncertainty, or in choosing appropriate criteria, that are troublesome. This lack of a guiding theory must be expected, for systems analysis is a relatively new discipline.

Systems analysis, particularly of the type required for military decisions, is still largely a form of art and not of science. An art can be taught in part, but not by means of definite fixed rules which need only be followed with exactness. Thus, in systems analysis we have to do some things that we think are right but that are not verifiable, that we cannot really justify, and that are never checked in the output of the work. Also, we must accept as inputs many relatively intangible factors derived from human judgment, and we must present answers to be used as a basis for other judgments. Whenever possible, this judgment is supplemented by inductive and numerical reasoning, but it is only judgment nonetheless.

One hope for guidance is to turn to science. The objective of systems analysis and operations research, in contrast to that of pure science, is primarily to recommend – or at least to suggest – policy, rather than merely to understand and predict. Thus systems analysis seem to be more nearly engineering than science. For purposes of distinction, one might say that science seeks to find things out while engineering uses the results of science to do things well and cheaply. Systems analysis has this latter objective; while every possible use of science and scientific methods is made, additional guidance is required, for it is necessary to decide what is well and what is cheap in each given situation.

Thus, systems analysis is sometimes described as the application of the 'scientific method' to problems of economic choice. Even though it is by no means clear that there is any unique method which can be termed scientific, the analysis advances through something like the following stages:

1. Formulation: clarifying, defining, and limiting the problem.
2. Search: determining the relevant data.
3. Explanation: building a model and exploring its consequences.
4. Interpretation: deriving the conclusions.
5. Verification: testing the conclusions by experiment.

A systems analysis always involves the first four of these stages but frequently omits the last. In military systems analyses, experiment is ordinarily not available; if we are lucky, our weapon system will be obsolete before there is a war, and we will never find out whether it was really satisfactory or not. The analysis, however, advances by iteration, passing through these stages more than once.

The discussion of method is divided into four sections, corresponding to the first four stages listed above.

Formulation

Formulation implies an attempt to isolate the questions involved, to define the meaning of the variables or factors that are operative, and to state relationships among these factors. The relationships may be extremely hypothetical, since empirical knowledge may be in short supply, but they will help to make the logical structure of the analysis clear. In a sense, this is the most important stage, for the time spent restating the problem in different ways, redefining it, or expressing its limits, brings to light whether it is spurious or trivial and points the way to its solution. The tendency all too frequently is to accept the original statement of what is wanted exactly as proposed, and then to set about building a model and gathering information, scarcely giving a thought to how the answer will contribute to the decisions which it is trying to assist. In fact, because the concern is with the future, the major job may be to decide what the policy maker should want to do. Since systems studies have resulted in rather important changes not only in how the policy maker carries out his activity but in the objectives themselves, it would be self-defeating to accept the customer's view of what the problem is.

An analogy with medical practice may be drawn. No doctor ignores a patient's description of his symptoms, but he cannot allow the patient's self-diagnosis to override his own professional judgment. The medical analogy is not entirely applicable, however

– the businessman or military commander ordinarily knows more than anyone else about his operations and what, if anything, might be wrong with them. Even so, he may not be so sound in his knowledge of how these operations affect, and are affected by, the context in which they occur.

How is the analyst to know his formulation of the problem is superior? *His one advantage lies in analysis.* That is, the process of problem formulation itself should be the subject of analysis. The systems analyst always has some idea as to the possible solutions of the problem; otherwise, he probably should not be working on it, for his analysis will prove to be too formal and abstract. At this early stage the analyst essentially makes an attempt to solve the problem before the facts are known. It is this attempt which gives him a basis for better formulation.

The problem itself does not remain static. Interplay between a growing understanding of the problem and of possible developments will redefine the problem itself. Primarily, as the result of discussion, the original effort to state the problem should suggest one or more possible solutions or hypotheses. As the study progresses, these original ideas are enriched and elaborated upon. Each hypothesis serves as a guide to later results – it tells us what we are looking for while we are looking. The final statement of the conclusions and recommendations usually rests on a knowledge of facts about the problem which are not known to the analyst at the start. Frequently, a hypothesis must be abandoned and an entirely new one considered.

Analysis must be an *iterative procedure*; that is, a process of successive approximation. The various stages in analysis – problem formulation, data collection, model building and computation – may follow each other in that order only within a single iteration; the tentative solution we obtain at the end of the first approximation helps us to better prepare a second formulation of the problem. Figure 1 attempts to indicate this iterative character of analysis. In a certain sense it is impossible to formulate a problem completely before it is solved, or, in other words, the final problem statement may have to be written simultaneously with the final answer. It is not a mistake to hold an idea in the early stages as to the solution; the pitfall is to refuse to abandon such an idea in the face of mounting evidence.

Even for small-scale individual problems, the number and complexity of factors under consideration at any one time must be reduced until what is left is manageable. In systems analysis, the complexity of the 'full' problem frequently far outruns analytic competence. To consider anything like the complete range of possible alternative solutions may be impossible. The vast majority of alternatives will be obviously inferior; there is no harm in leaving these out. The danger is that some solution which is better than that uncovered by the analysis will also have been left out. The number of alternatives available in completely unrestrained situations are too numerous to be examined. Constraints must be imposed, but by preliminary analysis, not arbitrary fiat. Such constraints must be regarded as flexible so that they may be weakened or removed if it appears in later approximations that their presence is a controlling factor.

Something must always be left out, otherwise problems are too big. For example, the decision to use a particular air-speed indicator in a new fighter should fundamentally rest on the military worth of the available indicators. It is futile to try to make this choice by considering all possible wars in which this equipment might be used. Yet, even though it may be beyond his

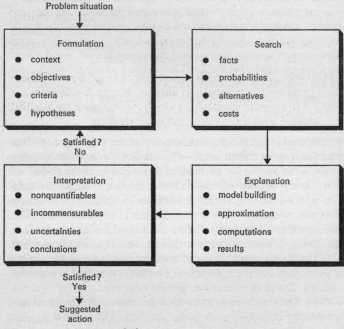

Figure 1 Activities in analysis

capability to do a complete job, the analyst can at least do some thinking about the larger problem. The dangerous path is to reduce the problem by fixing factors which would have been allowed to vary, if sufficient thought had been given to the larger problem.

Certain elements are common to systems analysis as well as to every problem of economic choice, although they may not always be explicitly identified:

1. *The objective* (or objectives). Systems analysis is undertaken primarily to suggest or recommend a course of action. This action has an aim or objective. Policies or strategies, forces or equipment are examined and compared on the basis of how well and cheaply they can accomplish this aim or objective.

2. *The alternatives.* The alternatives are the means by which objectives can be attained. They need not be obvious substitutes or perform the same specific function.[1]

3. *The costs.* Each alternative means of accomplishing the objectives involves certain dollar costs or specific resources used up.

4. *A model* (or models). This is a set of relationships, mathematical or logical, relevant to the problem at hand, by means of which the costs incurred and the extent to which the objectives are attained are associated with each alternative.

5. *A criterion.* This is the rule or test needed to tell how to choose one alternative in preference to another. For each alternative, it attempts to measure the extent to which the objectives are attained as compared with the costs or resources used up.

A characteristic of systems analysis is that the solutions are often found in a set of compromises which seek to balance and, where possible, to reconcile conflicting objectives and questions of value. It is more important to choose the 'right' objective than it is to make the 'right' choice between alternatives. The wrong objective means that the wrong problem is being solved. The choice of the wrong alternative may merely mean that something less than the 'best' system is being chosen. Frequently we must be satisfied with merely a demonstration that a suggested action is 'in the right direction', anyway. This may be all that is possible.

In the choice of objectives, the iterative character of systems analysis stands out. It is impossible to select satisfactory objec-

1. Thus, to protect civilians from air attack, warning shelters, 'shooting' defense, counterforce and retaliatory striking power are all alternatives.

tives without some idea of the difficulty and cost of attaining them. Such information can only come as part of the analysis itself.

The problems in the selection of suitable objectives and criteria are the most difficult in systems analysis. (See Hitch, 1958, 1960 and 1961, for detailed discussion).

The costs to be considered in choosing among alternatives, moreover, should be the 'new' costs, that is, the net additional resource drain or 'incremental cost' that would be incurred because of the choice of a particular alternative. Because a certain system may inherit facilities, personnel, or equipment from previous systems, its incremental costs may be much lower than what it would cost if it were to exist 'in isolation.' Also in comparing military capabilities, costs have sometimes been computed on the basis of what the various systems would cost independent of the existence of other systems or other capabilities. In this light consider, for example, a Navy supercarrier. In a paper comparison to estimate its value in a limited-war role, if no credit were assigned to its central war capabilities, then on a cost-effectiveness basis it would be handicapped unfairly in comparison with a weapon system that had only a single role.

Great attention must be paid to initial conditions; that is, to the assumptions which limit the problem and set the background against which the initial attempt at a solution is to be made. The situation is not like that of an empirical science, which starts with observed facts, but more like that of mathematics, where the results take any 'validity' they might have in the real world from the initial assumptions. The difference is that for the systems analysis to give correct guidance, it is important that the assumptions be the 'right' assumptions.

Once the problem has been broken down into its components – which is what analysing the problem means – some of the components can be further analysed, using various techniques; but others may defy analytic techniques. In that case, because the problem has been broken into smaller pieces, the systems analyst may be able to find individuals who have direct, sound experience and on whose 'considered' judgment he can rely.

Considered judgment differs from intuitive judgment in that the logic behind the opinion is made explicit. Both are based on an individual's experience and background, but when the reasoning is explicit, an observer can form his own opinion from the information presented. Judgment permeates systems analysis –

judgments as to which hypothesis is better than another, or which approach is more fruitful, or what facts are relevant. The ideal is to keep all judgments in plain view.

Uncertainty in long-range military planning problems being as great as it is, it is well – particularly early in the study – not to attach much significance to small differences in cost and effectiveness of alternative systems. Specifically, it is important to look for differences of the sort that have a chance of surviving *any* likely resolution of the uncertainties. The question to address is which alternatives have a clear advantage rather than the question as to precisely *how much* better one alternative is than another or even, initially, which ones move us toward the attainment of the objectives.

Search

This phase is concerned with finding the facts, or evidence, on which the analysis is based. It is necessary to look for ideas (and evidence to support them), including the invention of new alternatives, as well as to look for facts. Unless we have alternatives, and ideas about them, there is nothing to analyse or to choose between. If in the end we are to designate a preferred course of action, we must have discovered earlier that such a course exists. In long-range problems, the total number of alternatives may be endless, and we must use judgment to eliminate the unreasonable.

Many facts are hard to come by. The actual operational performance of future weapons in combat cannot be predicted with any degree of certainty. Purely theoretical studies or operations research of weapon characteristics must be depended upon. In systems analysis as contrasted with most other forms of engineering, a great many more inputs are a matter of judgment rather than a result of measurement or engineering analysis.

When should an inquiry stop? It is important to remember that in this sort of a problem, inquiry is rarely exhaustive. Inquiries are partial, and the decision maker must get along without the full advantage of all the potentiality of operations research and the scientific approach. Inquiries cost money and time; they cost in whatever values one is dealing in. They can cost lives; they can cost national security. It might be interesting to know what the Russians could do if we succeed in dropping an armed Atlas on Moscow. It might be an easy observation to make, but some of the costs seem to prohibit this type of investigation. One should never fall into the error of feeling that inquiry is free of cost. There are many contexts in which we can ignore the cost of inquiry; but paradoxes arise if we allow ourselves to forget that almost all inquiries must stop far, far short of completeness, either for lack of funds, or of time, or of justification for spending further funds or time on them. It is out of the question to collect all the information that is required for exhaustive analysis, and it is out of the question to process it.

As an analogy, consider the example of a physician who uses a clinical laboratory to help him decide whether or not his patient has one of several obscure ailments which have many similar symptoms. Even when all the reports are in, the doctor's inquiry may not be complete. He could probably do a lot more laboratory analysis. If the problem is simply one of diagnosis, one of the very best procedures would be to slaughter the patient and perform a thorough autopsy. The cost here is prohibitive, not only prohibitive by the standards of modern society but prohibitive simply by the fact that the physician's goal is to help the patient live a longer and fuller life. He would only frustrate himself if he bought knowledge at the price of the life he is trying to guard.

Explanation

After obtaining some idea of what the facts and alternatives are, it is necessary to build up some way to explain them and to determine their implications.

In order to make much progress with real-world problems, we must ignore a great many of the actual features of a question under study and abstract from the real situation certain aspects, hopefully, the relevant ones, which together make up an idealized version of the real situation. This idealization we call a 'model'.

In the general process of formulating a problem and gathering data about it, the analyst will have developed some ideas of what the major influencing factors are, that is, the factors which provide discrimination with respect to the possible courses of action. To produce quantitative results, it is necessary to assign a scale of measurement to each factor and to show its dependence on certain parameters. Next, the interaction of the factors must be described. Then we have a model. That is, the result of isolating those factors pertinent to the problem or the decision at hand, abstracting them, assigning a scale of measurement, and then describing their interactions builds the model.

For most phenomena, there are many possible representations;

the appropriate model depends as much *on the question being asked* as on the phenomena about which it is asked. There are thus no 'universal' models – that is to say, no one model that can handle all questions about a given activity.

Sometimes representation by the model is mathematical, by means of a series of equations or a computer program. At other times, particularly where detailed specification of the relationships between factors is extremely difficult – for example, in studying the behavior of human organization – the representation may be by a simulation or by a war game. A gaming model cannot be expected to tell us what an optimal response to an uncertain state of affairs might be, but it can do much to make the players aware of such uncertainties and of the necessity of formulating their plans in such a way as to cope with all foreseeable contingencies. Indeed, an important asset to all systems analysis is the spirit of gaming. This consists in explicitly looking at possible moves and countermoves, in examining and designing a wide range of alternatives, and in looking for substitution possibilities – all against a hostile opponent.

It should be emphasized that, in many important systems analyses, no need arises to build formal models explicitly. When such cases occur, the analysis may be extraordinarily effective since it may be completely understood by the policy maker. The essence of systems analysis is not mathematical techniques or procedures. A computing machine or a technique like linear programming may or may not be useful, depending on the problem and the extent of our information. The essential thing is a listing of the alternatives and an examination of their implications and costs in order that they may be compared.

The widely useful operations-research techniques for optimization, when they are used at all in systems analysis, are used much more extensively in component studies than they are at the heart of the over all problem. Before any mathematical technique can be applied to a real-world problem, we must construct a quantitative model of the processes involved. This model expresses the effectiveness of the alternatives under examination as a function of a set of variables some of which are under control. Once this is done, a solution can be determined mathematically, since formal statements of relationships between the variables exist. The solution obtained from such a model will be a usable solution to the real-world problem if and only if the model is a reasonably accurate representation of the real-world situation with respect to the question at issue. In situations of great complexity, such as those associated with major military decisions, only pieces of the problem can be represented with confidence. The sub-models for these pieces or components can frequently be put in a form in which they can be handled by techniques like dynamic programming or queueing theory. But even here, the new and more advanced techniques, while they are useful and promise to become more so, are seldom necessary since – except in relatively few instances – more elementary tools are usually adequate.

The design of models to assist in the decision process is in large measure an art, for it requires selection or composition, plus instinct and a sense of form, to achieve a desired effect. Wide experience and the collaboration of many people are helpful; but in cases in which we are modeling a complex future situation, modeling must be accepted as an art.

All of the assumptions of the model must be made explicit. If they are not, this is a defect. A mark of a good systems analyst (or any wise person communicating with others) is that he state the basis on which he operates. This does not imply necessarily that he makes better assumptions but only that his errors will be more evident.

The contrast between the relative amount of time usually spent on designing a model and that spent in computing its consequences can give bias in judging what is important. It is the design of the model and the faithfulness with which it represents those aspects of the phenomena being modeled which are significant for the question under consideration, not how far or how extensively we push the computation.

The validity of conceptual or mathematical models cannot, in the type of analysis we have been talking about, be tested by the methods of controlled experiment. The best that can be done is to test them by their workability. For example, we try to determine answers to the following questions:

1. Can the model describe correctly and clearly known facts and situations?
2. When the principal parameters involved are varied, do the results remain consistent and plausible?
3. Can it handle special cases in which we have some indication as to what the outcome should be?
4. Can it assign causes to known effects?

Whether or not one model is better than another does not depend on its complexity, realism, or computability, but solely on whether it gives better predictions.

'Working' the model, trying out various strategies and concepts of operation, is the nearest thing systems analysis has to scientific experimentation. Deductions based on operating the model frequently suggest new directions of effort. That is to say, starting with the relatively few parameters which characterize the systems in terms of the model, it is important first to show what changes in these would improve the performance of a system as measured by the model, and then to consider whether corresponding changes could also be made in the real system which would lead to improved performance in the real world. In this way, working the model contributes to system design.

Two aspects of model building are particularly troublesome: quantification and the treatment of uncertainty.

Some variables are difficult to quantify, either because they are not calculable, like the probability of war, or because no scale of measurement has been set up, like the effect on NATO solidarity of some unilateral United States action. This leads either to their neglect, for they tend to be ignored, or to their entry only through a qualitative modification of a solution achieved through the manipulation of variables which have been quantified. Thus the effect of the quantitative variables is built in, while the nonquantitative ones are subject to forgetfulness and may be easily lost in the welter of qualitative considerations that must be weighed, when the problem of what action to recommend on the basis of the solution from the model arises.

One argument for the omission of a particular variable is that the solution of the problem is virtually insensitive to it. The fact that many variables fall into this category makes analysis possible. If the results were not insensitive to all but a relatively small number of variables, analysis would have to yield completely to guesses and intuition. Insensitivity can occur either because a factor is irrelevant or trivial in its quantitative effects or because it has roughly the same effect on all of the alternatives under consideration. *The point is that this insensitivity must be discovered.* Sometimes logical reconnoitering is sufficient, but usually analysis is required, with arbitrary values assigned to factors we are unable to calculate.

If nonquantitative variables are not to be neglected without mention or dismissed with some spurious argument, such as the one that they act in opposite directions and hence cancel out, then how are they to be treated? The usual method is to attempt to take them into account through modification of the solution rather than to incorporate them into the model. But this in itself represents a particular method of quantification, for, by altering the solution to take account of the previously omitted variables, the analyst is implicitly valuing them. Since we always have some insight into the range of values that a factor might take, we can, even in the worst cases, assign the factors an arbitrary value and observe the effect on the solution. It seems to be an empirical fact that results seldom come out of optimization problems until they are quantitative; consequently, every effort should be made to quantify.

Systems analysis is concerned with problems in which the essence is uncertainty about the future. Such analysis, as well as any other attempt to answer the same questions, must necessarily face this uncertainty squarely, treat it as an important element in the problem, and take it into account in formulating recommendations. The treatment of uncertainty is not merely a difficulty in principle, but is a considerable practical problem.

Statistical uncertainty – that is to say, uncertainties having more or less objective or calculable probability of occurrence – can be handled in the model by Monte Carlo or other methods. Such uncertainties, like those in cost or missile accuracy, can be annoying, but not devastating, like the 'real' uncertainties associated with the prediction of what the environment may turn out to be during the lifetime of the systems under consideration.

These latter uncertainties about the future behavior of things that are beyond the practical ability of analysts to predict belong to the class of real uncertainties. Under real uncertainty, we consider events to which individuals may attach subjective probabilities, like the probability of war, but which we cannot calculate. With regard to air defense, for example, real uncertainty involves such questions as, 'Will we have warning? If we get it, will we believe it? What surprises does the enemy have?' For such uncertainties, there is frequently widespread disagreement about the pertinent probabilities and even confusion and vagueness within any one person.

The best way to compensate for uncertainty is to 'invent' a better system or policy which provides insurance against the whole range of possible catastrophes; the difficulty is to discover how to do this.

Sensitivity and contingency analyses help us to select or

design the alternatives so that their performance will not be sensitive functions of unknown parameters.

In 'sensitivity analysis' several levels are used for values of key parameters – not just the expected or most probable values – in an attempt to see how sensitive the results are to variations in these parameters. The hope is to obtain a dominant solution in which the ranking of the preferred alternative is essentially insensitive to reasonable variations in values of the parameters in question. 'Contingency analysis' investigates how a system chosen with one assumption about the environment would measure up to the performance of its alternatives, if radical changes in the environment were to occur. Thus, sensitivity analysis might test the alternatives for a wide range of enemy capabilities or for the consequences of having planned for one level of capability when another is experienced. Contingency analysis might test the alternatives under a change in criteria or compare them in an environment in which France, say, had become part of the Communist Bloc.

Since a system analysis is a study which attempts to influence policy, it must make a convincing comparison of the relevant alternatives. It must demonstrate that some course of action, A, is better than alternative possible courses of action B, C, D. . . . To do this, the analysis may have to be done in two stages: first, find out what to recommend, and second, make these recommendations convincing.

After we are convinced to our own satisfaction what the preferred system or policy is, how can we show that under any reasonable assumption the system or policy designed or selected by the analyst is indeed to be preferred? One way to do this is to use either an *a fortiori* or a break-even analysis. To make an analysis *a fortiori*, we bend over backwards in making the comparisons to 'hurt' the system we think is best and 'help' the alternative systems. If it then turns out that after we have done this we can still say we prefer the handicapped system, we are in a strengthened position to make recommendations. Sometimes we cannot do this – say, if we concede the exaggerated performance claims for rival systems and the pessimistic estimates about the systems we like. In this case, we might try a *break-even* analysis: we find what assumptions we have to make about important values in order to make the performance of the two systems come out to be essentially the same. Then we can simply ask people to judge whether these assumptions are optimistic or pessimistic.

Interpretation

After a solution has been obtained from a model, this solution must be interpreted in the light of considerations which may not have been adequately treated by the model, since the model was but a single representation of the real world chosen by the analyst. The solution of a problem which has been simplified and reduced to mathematical form by drastic idealization and aggregation of the real-world factors is not necessarily a good solution of the original problem.

In the attempt to interpet the results of analysis, there are special problems associated with military questions. Many factors used in the computations are not and cannot be measured. Sometimes this is because of time limitations; other times it is because factors such as the enemy defense strength, or degradation in combat of complicated man-machine combinations, are not accessible to measurement but have to be assessed on the basis of experience or pooled judgment. The results of computations must be examined to see if they depend critically upon estimations such as these.

It is important for the man who is to use analysis to distinguish between what the analysis shows and the recommendations for action the analyst makes on the basis of what he thinks the study implies. Frequently, when new minds – management, for example – review the problem, they bring new information. Even though the solution obtained from the model is not changed, recommendations for action based on it may be.

Practices such as that suggested by the following statement can lead to serious error: 'If several alternatives have similar cost and effectiveness, and these results are quite sensitive to the values assigned to the input, some other basis for decision must be found.' This may amount to saying that if, after honest analysis, it must be concluded that we are fundamentally uncertain which of several alternatives is best, the issue should then be resolved on the basis of some specious side criterion not originally judged adequate to discriminate. On the contrary, the points to stress, if such results are found, is that the decisions must be made on the basis of forthright recognition of the fundamental uncertainty. The implication is that in this case unique optimization results are not to be trusted, and therefore they should not be trusted.

If, in the judgment of the analyst and those who are to use his analysis, the alternative ranked highest is good enough, the process is over; if not, more and better alternatives must be designed or the objectives must be lowered. Analysis is sufficient to reach a policy conclusion only when the objectives are agreed upon by the policy makers. In defense policy in particular, and in many other as well, objectives are not, in fact, agreed upon. In these cases, the choice, while ostensibly between alternatives, is really between objectives or ends. Hence, non-analytical methods must be used for a final reconciliation of views. The consequences computed from the model may provide guidance in deciding which objectives to compromise. It is not obvious how to do this, however, and judgment must again be applied.

By definition, no judgment is known to be correct. Because systems analysis ordinarily goes beyond objective analysis, it relies heavily on considered judgment. No matter what may be the hopes of professional analysts, the judgment applied by the decision maker in the last phase of a study limits the influence of the previous analyses. At its best, analysis can embrace only a part of a broadscope problem, it gets no foothold at all on some objective elements, and before it organizes an understanding of all objective elements it becomes too complex to handle.

Concluding comments

The past few years have seen marked changes in the application of systems analysis to military problems. In the words of R. D. Specht (1960), RAND analyst:

Let me put the differences inaccurately but graphically: in our youth we looked more scientific. That is to say, we attached more importance, years ago, to the business of representing that part of the real world with which we were dealing by a single analytical model. With the context chosen, the assumptions determined, the criterion selected, we could turn our attention to the more intriguing questions of how best to apply modern mathematical techniques and high-speed computers to produce a neat solution from which conclusions and recommendations could be drawn.

There are many problems in the world for which this is a sensible, even a recommended approach. There are problems impossible of solution without the use of the most powerful tools of mathematics and of computers. The optimal distribution of weight and thrust between the several stages of a lunar probe, the determination of its initial trajectory – these are well-defined questions and yield to neat and orderly solution. On the other hand, the stability of the thermonuclear balance or the composition of a strategic deterrent force or the character of the next generation of tactical weapons – these are not questions that may be attacked usefully in this manner, although essential fragments of these problems may be solved analytically. A trivial reason for this is that even modern techniques of analysis are not sufficiently powerful to treat these problems without brutal simplification and idealization. The major reason, however, for the inadequacy of simply optimization procedures is the central role that uncertainty plays in this sinful but fascinating world. No longer are we analysing a problem with a given and definite context and with specific equipment. We may not have clearly defined objectives. Instead, we must try to design – not analyse – a system that will operate satisfactorily, in some sense, under a variety of contingencies that may arise in a future seen only dimly.

We have learned that new tools – high-speed computers, war gaming, game theory, linear and dynamic programming, Monte Carlo and others – often find important application, that they are often powerful aids to intuition and understanding. Nevertheless, we have learned to be more interested in the real world than in the idealized model we prepare for analysis – more interested in the practical problem that demands solution than in the intellectual and mechanical gadgets we use in the solution.

The analytic method, in contrast to many of its alternatives, provides its answers by processes which are reproducible, accessible to critical examination, and readily modified as new information becomes available. At the very least, systems analysis can supply a means of choosing the numerical quantities related to the weapon system in such a manner that they are logically consistent with each other, with the general objectives of warfare, and with the calculator's expectation of the future. Systems analysis must be tempered with and used alongside experience, judgment and intuition. It cannot replace these other approaches, but it can help build a framework in which they can operate more efficiently.

References

HALL, A. D. (1962), *A Methodology for Systems Engineering*, Van Nostrand.
HITCH, Charles (1958), 'Economics and military operations research',
    Rev. of Econ. and Stats, August, pp. 199–209.
HITCH, C. J. (1961), 'On the choice of objectives in systems studies', in
    Eckman, D. P. (ed.), *Systems: Research and Design*, J. Wiley
HITCH, C. J., and MCKEAN, R. N. (1960), *The Economics of Defense
    in the Nuclear Age*, Harvard University Press.
MCKEAN, R. N. (1958), *Efficiency in Government Through Systems
    Analysis*, J. Wiley.
SPECHT, R. D. (1960), 'RAND: a personal view of its history'
    Ops. Res, vol. 8, no. 6, November–December.

8 S. P. Nikoranov

Systems Analysis: A Stage in the Development of
the Methodology of Problem Solving in the USA

Excerpt from Stanford L. Optner, *Systems Analysis for Problem Solving*,
Prentice-Hall, 1965.[1]

The problems of selecting armaments for the Army, Air Force
and Navy, selecting the most profitable production, selecting the
direction for research, analysing problems of urban development
including urban transportation, determining a rational policy
with respect to natural resources, in particular water resources . . .,
these and similar problems began to acquire a substantially
novel character in the USA in the 1940s and 1950s. As the
problems grew in size, some of them, such as communications
with the aid of satellites, acquired global proportions.

The complexity and difficulty of managerial problems also
increased sharply with time. The interdependence of questions,
which formerly had seemed unrelated, became increasingly mani-
fest. The urgency to solve problems increased considerably.
Expenditures to implement particularly [complex] solutions could
require tens, hundreds of millions or even billions of dollars,
while the risk of failure in problem solving became ever more
palpable. It therefore became necessary to take into account an
ever-increasing number of interrelated conditions, while the time
for problem-solving was rapidly running out.

The basic question underlying the solution of any given problem, irrespective of its content, character or general area, was
the selection of the most suitable alternative course of action.
The choosing of alternatives depended on the ability to estimate
the effectiveness of each alternative, and to estimate the necessary
cost to develop and implement it. Similar analytical attitudes had
been adopted in the areas of capital investment and industrial
expansion before the Second World War.

Formerly work on the development of a system of weaponry

1. Written in 1969 as an introduction to the Russian translation. The
Russian language introduction was translated into English in 1970 by
Dr Ilija Yoksimovich, Professor of International Relations, Shaw
University, Raleigh, North Carolina. The Russian translation was edited
by S. L. Optner for presentation in this form.

was begun without research as to how the system in question might be used, what the cost would be and whether or not its contribution to defense would justify the expenditures involved in its creation. This happened because prior to the Second World War, arms expenditures were relatively small, there were few possibilities of choice and the operative principle was 'nothing but the best'. During the Second World War, and with the onset of the atomic age, the costs of arms development increased many times, and this approach became unacceptable. It was gradually replaced by another [principle]: only the absolutely necessary, and that for the least possible cost. However, to make it operative, the new principle required techniques to identify, evaluate and compare alternative solutions.

Mathematical methods used in industry and commerce, as well as computer models for the investigation of operations, which had been developed by this time, could not be used in view of limitations inherent in them. Methods were required which would permit the analysis of complex systems as a whole. New methods would ensure consideration of many alternatives, each of which would be describable in terms of a large number of variables. These methods would also be required to encompass the completeness of each alternative, help to introduce measurability and permit an assessment of uncertainties.

The broad and universal problem solving methodology which emerged as a result of these needs was called systems analysis. The new methodology, created for the solution of military problems, was first utilized exclusively in this area. However, it soon became clear that civilian problems, business problems, financial problems and many others not only permitted but actually required the use of this same methodology.

The wide utilization of systems analysis led to its improvement and perfection. The tendency, characteristic of the USA, to give everything a 'commercial product' or 'goods' aspect, led ultimately to its widespread acceptance. Systems analysis quickly absorbed the achievements of many related, parallel disciplines, and became a scientific and applied discipline in its own right. It was rich in forms and applications, had a unique character and direction, and was practiced as an area of professional creativity. Since systems analysis received continuous application in the solution of problems, it soon began to exert a deep influence on the understanding and practice of management problem solving and business in general.

The Soviet reader is still ill-acquainted with the history and content of systems analysis. Yet, a familiarity with systems analysis, with its applications and its achievements, is of great inherent interest. The study of the objective foundations of systems analysis, of its general approach and of its particular methods might prove to be extremely useful. [It might be used] to probe some aspects of the methodology of long-range planning in the areas of national economy, in the choice of directions for technological development, in the creation of mechanized systems of navigation, in the solution of problems of organization of scientific research and in the development of technology.

A study of systems analysis demands a certain amount of concentration of the Soviet reader to shift the objective elements of the methodology from its socially-determined form in which it is used in American literature and practice. The solution of problems is required in every socio-economic organization of society. However, the concrete forms of problem definition and content, the causes of their emergence, the forms of solution, the organization of solution-content . . ., all these depend wholly upon the character of the socio-economic order. Problems of the business world and industry are, in the last analysis, the problems of a developed, industrial nation. The forms which problem-solving take are determined by organizations controlling national institutions: for example, the Department of Defense and private enterprise. Therefore, the study of systems analysis and the history of its development may be of definite interest in understanding some aspects of scientific and social change in the USA.

In order to help the reader orient himself to the range of questions relating to systems analysis, the following will be briefly considered: the basic concepts and ideas of systems analysis; the contribution of systems analysis to the methodology of problem solving and the improvement of business organization; the history of the development and the applications of systems analysis; and the possible direction of its future development.

Basic concepts

The simplest way to understand systems analysis is to describe its most basic concepts and assertions. Systems analysis is a methodology for solving major problems based on the notion of systems. Systems analysis may also be regarded as the methodology of operating the business enterprise, since businesses are the entities which use the problem solving methodology. Both of these definitions are indissolubly linked with each other. However to begin with, we shall analyse the problem solving methodology as such, and only later [suggest] its influence on business.[2]

At the heart of the methodology of systems analysis is the notion of quantitative and/or qualitative comparisons among alternatives. The explicit aim is the selection of one alternative to be attained or adopted. If a criterion of qualitative excellence is met, as between alternatives, then quantitative estimates may be obtained. In order that quantitative estimates permit a comparison between alternatives, they must reflect those properties of the alternatives which take part in the comparison. This may be attained if all the elements of a particular alternative are taken into account, and if estimates are made for each element.

Hence, there arises the notion of focusing attention upon all the elements necessary to a given alternative. In layman's language, this may be expressed as consideration of every side of the question. The generic whole which emerges from this definition is what is known in systems analysis as the total system, or simply the system. A system, therefore, is that which solves the problem.

But, how is this generic whole, this system, to be isolated? How is one to determine whether a given element is or is not a part of a given alternative? The only possible criterion is [a knowledge of] the role of each element in a process leading to the emergence of the end result of the given alternative. And if this is fundamentally true, then the nature of process turns out to be the central [issue] in systems analysis. The first thing which must be isolated, if we are to think and act systematically, is process. Systems thinking is impossible without an understanding of the term *process*.

A system is determined by a given set of system objects, properties and their relationships. The system objects are input, process, output, feedback and a restriction. Input is that which [energizes and] which changes the operation of a given process. In many cases, the components of input consist of the 'passive input' [of the entity] which is doing the processing. The end result of the final state of the process is known as the output. The interconnection, or relationship between systems determines the continuation of processes: i.e. the output of one process may be the input to some other. Each input of a system is also

2. In this [article] we shall follow S. L. Optner and S. Young, primarily.

the output of some other system, and conversely. To isolate a system in the real world means to indicate all processes which yield a given output. The most unpredictable systems are human creations: i.e. systems which are the output of consciously executed processes of a human being.

In each system, there are three separate subprocesses distinguishable by the respective roles they play in the whole: the basic process, feedback and restriction. The basic process transforms input into output. Feedback performs a number of operations: it compares the actual output with an objective (a model output) and identifies the differences; it evaluates the content and significance of this difference; it works out the solution, both the coincident and the disparate; it formulates the means of introducing [solution data as] additional input; and it interacts with the basic process with the aim of achieving the objective, using the actual output.

The restriction is initiated by the purchaser[3] of the output system. He is the one who utilizes the output. This restriction limits the products of the system, conditioning them so they may ultimately correspond to the demands of the user. The restriction consists of the aims (objectives) of the system and its constraints, the limiting conditions placed upon the objective. The constraints should be consistent with the aims.

Every system consists of subsystems. It is postulated that every system may be described in terms of system objects, properties and relationships. The boundary of the system is determined by the separation between the totality of inputs, processes and outputs [required to operate a given system], and the surrounding environment. The surrounding environment is the totality of natural and artificial systems.

A problem situation is characterized by the difference between the necessary (desirable) output and the existing output. An output is deemed necessary if its absence constitutes a threat to the stable, predictable performance of the given system. An existing system output is produced by an existing system. A desired system output is postulated by [proposing] a desired system. A problem may consist in the prevention, diminution or increase of output, or else in the change of output. The actual conditions surrounding the problem represent the existing system (the

3. In the sense of the manager who underwrites the one-time costs of system design and implementation. [Ed.]

'known'). [Systems requirements] represent the [aims of the] desired system. The solution of the problem (the 'unknown') closes the gap between the existing and the desired system.

Problems may be recognized through [the evaluation of output]. Discovery of a problem is the result of a process of [continuous monitoring of a system]. Identification is possible given a knowledge of norms, or of the desired behavior of the system.

The process of finding solutions is concentrated around iteratively performed system operations. If the relationships and the elements of a given problem are known, their identification is characterized by the determination of quantitative relationships, and the problem is termed quantitative. If the relationships and the elements of a [problem] are known only partially, identification may take on a qualitative character, and the problem is termed qualitative or ill-structured.

As the methodology of problem solving, systems analysis points out the axiomatically necessary serial relationship of interrelated operations; in the most general terms these consist of the isolation of the problem, the design of the selected solution, and finally the implementation of this solution. The solution process itself consists in the development, evaluation and selection of alternative systems, based upon criteria of cost, time, effectiveness and risk, taking into account the relationships among the limiting values.

The selection of boundaries for a given process is determined by the conditions, aims and the possibilities for implementation of a given process. The optimal use of this procedure presupposes a many-sided utilization of heuristic conclusions within the framework of a postulated structure of system methodology.

Reduction in the number of variables is performed on the basis of an analysis of the system's sensitivity to a change of variables. The collection of individual variables into aggregates, or 'factors', is performed by a suitable choice of criteria, and also wherever possible, by use of mathematical methods. The logical completeness of a process is guaranteed either by explicit or implicit assumptions, each of which may constitute a source of risk.

It is postulated that the objects of the system, and the ingredients of a solution of the problem, are standard for all systems and all problems. The only things which may change are the quantitative relationships among system objects. Improvements in method, at a given state of scientific knowledge, have a limit defined as a potentially attainable level. As a result of the solution of a problem, new system relationships are established, a part of which determines the desired output, while the other part deals with the unforeseen possibilities which become sources of future problems.

Such, in general terms, are the main concepts of systems analysis seen as a problem-solving methodology. Practical applications of systems analysis may be realized in two types of interaction: first, when a new problem originates; and second, when a new solution emerges, given that it was discovered without any immediate connection to a given group of problems.

The solution to a problem in a new situation is performed in the following pattern:

1. Discovery of the problem.
2. Assessment of its relevance.
3. Definition of aims and constraints.
4. Definition of criteria.
5. Determination of the structure of the existing system.
6. Determination of the defective elements [of the existing system] which hinder the attainment of an assigned objective.
7. Assessment of their relative importance with respect to the outputs of the system, as determined by criteria.
8. Determination of the structure necessary for a choice of alternatives.
9. Determination of the solution-finding process.
10. Selection of alternatives with a view to solution-finding.
11. Construction of a model to choose among alternatives.
12. Realization of the solution.
13. Evaluation of the consistency of the solution with the original objectives and constraints.
14. Evaluation of the results stemming from the realization of solution.

Systems analysis in business

Let us consider in what way systems analysis can be identified with business. To this end, it is necessary to regard companies not as a structure of dependent established relationships, but as a place for problem solving. This approach permits one to regard business as a system, and to use the conceptual apparatus of systems analysis to describe, study and improve company operations. There is an assumption that the unsatisfactory performance of a product implies the unsatisfactory performance of the business which makes the product. It follows, that an improvement in performance may be attained by means of an improvement in problem solving by means of systems analysis.

In order to improve the performance of problem solving, various methods may be employed, from a change in procedural documentation to the utilization of mathematical models and computers. The methods themselves may have alternatives, and selection of the best alternative may proceed according to the principles of systems analysis. The influence of all the individual efforts which in sum comprise the system analysis, from problem identification to solution realization, must be approximately the same.[4] It is meaningless to have powerful methods for obtaining solutions if problem identification is unsatisfactorily performed. For example, decisions as to the excellence of problem solving of an organization must emerge from an appraisal of the problems being analysed, and must ultimately be measured by their innate complexity. Particular methods for improving problem solving functions are only of significance if the organiza- tion adopts the position that business must be analysed as a complete system.

In the past, the role of the scientific method in problem solving was substantially circumscribed due either to the weakness or the total absence of, system concepts. Therefore, the need to view the organization as a system could not arise. Now, more efficient methods have been created for understanding and apply- ing the methods of problem solving and their development continues to be intensively and purposefully researched.

The application of individual unstructured methods [of problem solving] within the framework of existing organizations is difficult and ineffective. One reason for this is that the utilization of a method requires the isolation of a function as a complete process separate from the connected dependent causally related operations in which the function traditionally occurs. This is impossible without a change in the operations technique of the existing organizations. In organizations which are operated by essentially heuristic methods, and are not managed systematically, and in which traditional techniques dominate logical methods, functions are seldom isolated in such a way as to lend themselves to manipulation by powerful [problem solving] methods.

4. It might be more accurate to say that the effort invested should be proportional to the inherent difficulty and the state of the art in each particular problem, as well as their potential benefit to be realized. [Ed.]

Another reason for the failure of problem-solving methods may lie in the bureaucratic character of existing organizations. Methods of problem solving become substantially more complex if one has in mind not one method, but an aggregate method; and not one particular function, but a complex of mutually interrelated functions. An existing organization may well solve some problems with the traditional methods available to it, but it cannot effectively utilize contemporary scientific facilities in problem solving if it remains tradition bound.

When the gasoline engine was invented, it was placed upon a wooden carriage. As the engine was improved, the carriage was concurrently changed, until the modern automobile emerged. But a modern engine of 300 horsepower can no longer be placed upon a wooden carriage. The case for the powerful new methods (rules of problem-solution) is analogous: that is, such advanced techniques such as transportation models, queuing models, network models do not belong in the framework of tradition-bound organizations. It follows that modern organizations must be structured around new methods of problem-solving.

For the evaluation of alternative business systems, the analyst uses the criteria of measurability, effectiveness, reliability, optimization and stability. *Measurability* is the system's capacity to evaluate its own characteristics. If a system does not possess measurability, it is impossible to determine whether it has yielded an improvement or a worsening. *Effectiveness* is the analyst's ability to solve a problem with the aid of the given system. Effectiveness presupposes a balance between the parts of the system. It is meaningless to be in possession of a solution which is, at the same time, optimal, but not measurable and not effective. The solution must be measurable, effective and reliable before it is optimal.

The task of the top management of the organization is not the attainment of solutions, but the construction of a process for the attainment of solutions, and the observation of its functioning. The ability of an average manager to propose good solutions is not a criterion for his promotion to higher management. To do this would be analogous to relegating the research on a weight-lifting machine to a weight lifter, solely upon the basis of his ability to lift weights.

Management which is concerned with day-to-day operation of large-scale problems and with the design and implementation of massive systems is called systems management. The complex of tasks within a particular problem is called a program. Hence, system management is sometimes termed program management.

By defining new problem-solving methods, using the framework of systems, systems analysis transferred to business the techniques which were well known in technological research and development. The systems approach has given the character of a research and engineering activity to the solution of business problems. Some scientists believe that the reorganization of business in accordance with the demands of systems analysis will lead to important changes in management within the coming decade.

The basic content of systems analysis does not lie in some formal mathematical apparatus describing 'systems' and 'problem-solutions,' although attempts for the creation of such an apparatus exist.[5] Nor do the solution processes lie in specialized mathematical methods, such as the formal calculation of uncertainty (although some work has also been done in this direction). It lies rather in its conceptual apparatus, its ideas, its approach and its constructs, described briefly in the following quote[6]:

In mathematics and physics certain axiomatic theories find wide application, and these are sometimes called *formal* or *deductive*. The axiomatic method is the most important of all theoretical investigations at the present time. This method is also beginning to be used in other areas of knowledge. Many of the achievements of mathematics, physics and of a number of other natural disciplines owe much to the axiomatic method. This has become possible for the reason that it permits retention of at least a part of the discipline in question as an integrated whole, in the use of broad logical structures. Axiomatic theories and their cause-finding methodologies are based on a number of postulated situations. They are, in the last analysis, reflections or generalizations of empirical experience.

Systems analysis is a *normative* methodology. A normative theory is analogous to an analytical theory. It does not predict what is going to happen in the given case, but it does predict what will happen if all the conditions and presuppositions of the relevant theory are met. Correspondingly, a normative methodology does not state what should be done in a given situation in order to obtain a given result, but it does state what should be done in the given case if all the conditions and presuppositions of the methodology are met.

Usually, normative theories are expressed in a mathematical form,

5. Charles J. Hitch (1965) p. 98.
6. Charles J. Hitch and Roland N. McKean (1960).

while normative methodologies are expressed in the form of algorithms. Although such forms may sometimes be adaptable to systems analysis, in the majority of work of an applied kind, a simplified interpretation of normative methodology is given. However, the non-mathematical form of expression should not serve to confuse the relatively normative character of the methodology.

The success of a normative methodology in application depends on the skill in the interpretation of its demands in practical situations. The more powerful a normative methodology is, and the more general the concepts used by it, the broader becomes the problem area for which solution the methodology may be used. The more difficult the determination of the limits of the methodology in accordance with its normative character, the greater its generality. Consequently, greater skill is required in its practical application. In order to be useful, a methodology must be constructed on the basis of the theory which is adequate to the objective and the task. For complex objectives and tasks, a methodology with a complex structure has to be developed. This circumstance also raises the requirement for the capability to use a normative methodology in these circumstances.

The application of normative theories to isolate elements of the problem-solving process is not new. Models utilizing operations analysis, such as models of game theory, are all normative theories. The appearance of these models stems from the use of axiomatic methods of mathematics and physics in the areas of operations.

Indeed a normative definition of the solution process is not new.[7] However, the wide, practical applicability of such a definition apparently must be regarded as an original achievement of systems analysis. Another achievement of systems analysis, one which is logically co-dependent with the construction of the methodology, is the transformation of the normative approach into the area of business operations.

The application of a normative methodology of such great proportion and character is a matter not to be taken lightly. The authors of written works on systems analysis emphasize, rather than minimize, the normative character of this methodology:

If you wish to increase the output of an organization, it must be regarded as a total normative system. Although this requirement is generally accepted on faith, nevertheless there exist several empirical

7. As an example see G. Polya (1957).

verifications (such as in the area of armaments) which confirm that this approach allows the user to obtain the best results. At present, (scientists and businessmen are) at the initial stages of development of this approach, and may only approximately describe the basic characteristics of total systems.[8]

Systems analysis is based upon concepts of a high level of generality and partially upon the following ideas: constraint, property [characteristic], process, quality, cognition etc. The logical structure of systems analysis is well developed. It includes of necessity some aspects of the cognition of reality. This is as it should be in a methodology of problem solving which includes the need to interact with reality.

Because of the formal, scientific burden of the methodology, one may expect that the application of systems analysis imposes weighty demands upon management and industrial groups, if they wish to make use of it in the solution of problems. But this is apparently already the case:[9]

Systems analysis may only be usefully applied, it seems, wherever there already exists environment of sophisticated management.

With respect to the normative methodology, systems analysis defines a basic nomenclature for functions which must be identified in problem solving; such as, it postulates a process which is fulfilled by the organization. This is in contradistinction to the traditional approach to organization which postulates an administrative system.

The postulated process lends a constructive character of the highest order to systems analysis. At the same time, it is difficult to avoid the impression that systems analysis in its present form, and even taking into account certain fully determined emerging tendencies, is in a certain sense decidedly one-sided. The adoption of process as the key concept leaves unresolved the empirical fact that processes exist only as relatively specialized totalities. This is implicitly considered by introducing the notion of limitations [constraints] upon the problem and the system. It is also handled by using the notion of open and closed systems borrowed from general systems theory. However, this does not compensate for the absence of a normatively introduced description of relatively specialized totality (such as that of the

8. For example, the work of A. C. Enthoven in Hitch and McKean (1960).
9. E. Jantsch (1967).

business firm), although constraints upon the methodology make it more practical.

Of great significance is the change in the thought patterns of individuals who have mastered the concepts of systems analysis. The ideas of systems analysis permit the user to separate the form of organization from its content. The former owes its character to its history; the latter remains the same for all methods. A mastery of systems methodology also gives the user an understanding of what *must* be . . . a comprehension of the necessity for [mixed] team work[10] an understanding of the desirability for improving the working methods of the organization, and a recognition of the analyst's role in the work process. Persons capable of 'system' thinking are in a position to reach agreement with respect to changes in organization by use of the new methods.

The discipline in the thinking of a group of people, dictated by a powerful and clear technique, is perhaps a more important matter than some mathematical methods (which might permit one to perform some calculations or operate a particular model). Within such an intellectually disciplined group, a norm of professional behavior becomes imperative; in other contexts, this attitude is usually something which is taken for granted, but which seldom materializes. The performance of individuals in the group, established by the management in the expectation of its own internal discipline and logic, fixes the methodological principles and the rules of analytical procedure. It permits an easy way to expose and remove disloyalty, incompetence, dishonesty and disorderliness.[11] At the same time, it becomes possible to clarify the reasons for disagreement because all elements of the solution process, including the positions held by those participating in argument, become demonstrable. The name given to a paper read before the US Senate by one of the leading specialists in systems analysis, which described the above concept was titled, 'Systems Analysis – Ground Rules for Constructive Debate.'[12]

Mastery of the methodology of systems analysis makes the following situation practically inevitable: the management consciously demands development of improved methods for a better

10. I have inserted the term *mixed* to indicate the use of individuals from different disciplines and specialities in a single project. [Ed.]

11. I regret the lack of a reference at this point. [Ed.]

12. A. C. Enthoven (1968) pp. 33–40.

operation of the organization; scientists develop those methods under a realistic understanding of the content of their assignment; management and personnel become active in the adaptation of these methods and make possible an increasing competence in their utilization.

In the instance of systems analysis, system methodology demonstrated its practical efficacy for the first time. But the development of systems analysis also had an important conceptual significance. The majority of scientific and applied areas of knowledge were integrated in the solution of problems. Among these were a few which had as their subject problem solving, itself. Systems analysis for the first time provided a generalized methodology of problem solving based upon the concept of a system.

Historical development of systems analysis

The leaders of systems analysis have not yet summarized the results of the nearly thirty-year development of their subject (if the very limited material of a lecture by Charles Hitch be discounted). However, existing data in the literature permits a compilation of general ideas concerning the historical development of systems analysis, even though these might prove to be debatable. We shall mainly consider the development of applications and the literature in the following material.

It is hard to say when systems analysis may have first been used. Goode and Machol (1957) point out:

For more than a decade, engineers and administrators have witnessed the emergence of a broadening approach to the problem of designing equipment. This phenomenon has been poorly understood and loosely described. It has been called *system design, system analysis* and often the *systems approach*. Rarely does the speaker using these terms intend to convey those concepts which are brought to the minds of his hearers, nor for that matter are any two hearers likely to be in agreement.

Soon after the start of the Second World War, the US Air Force submitted a request to the Harvard Graduate School of Business Administration to find a method of increasing, within one year, the then-existing Air Force strength of 4000 fighter aircraft and 300,000 personnel, to 80,000 aircraft and 2·5 million personnel, but in such a way as not to exceed a cost of $10 billion. In order to cope with this assignment, Harvard Business School organized a

Statistical Control Project. Participants included Robert MacNamara and a number of other company administrators. In approximately one year, the problem set by the Air Force was solved, and the arsenal of the military command was enriched by the method of 'systems analysis'.

Initial impetus: national defense

The initial development of application and popularization of systems analysis is due to the work of The RAND Corporation and other government-sponsored organizations. Theoreticians and specialists of these corporations completed a number of fundamental projects and systems analyses which are recorded in an immense bibliography. In addition, RAND and other companies provided the training ground for a number of people who were destined to put systems analysis to practical use in the Department of Defense and other organizations. Historical perspective may reveal that the most important contributions of the individuals and companies involved may not have been in the research projects themselves, but in the momentum their efforts gave to the development of systems analysis.

The RAND Corporation was established in 1947. In 1948, a group was organized within RAND for the evaluation of weapons systems. This group was to play a key role in the development and utilization of systems analysis. In 1950, a department was formed by RAND for the analysis of weapons costs, and to carry out research on the budgetary aspects of systems analysis. Early in the 1950s, the concepts of 'weapon systems', and the practice of 'systems management' began to be terms of common usage. The development of the B-52 supersonic bomber, which was begun in 1952, was among the first end products to be called a 'system'. The development of strategic missile systems, and of systems designs to repel air attack, were a great influence on the formulation of the techniques of systems analysis in the 1950s.

As diverse specialists and research organizations carried out investigations with the aid of the systems analysis, it became clear that its effective use could only be realized within a framework of a business structure which would sanction the use of the methodology. The methodology itself was already worked out in detail and articulated in a book which appeared in 1960, by a group of RAND specialists (see Hitch and McKean). At the same time, it became clear that centralization of management in the

Department of Defense according to military tasks, rather than according to type of military service (Army, Navy, Air Force), would fulfill its aims only if the programmed management and powerful methods of problem solving were to be utilized as the control mechanisms of the armaments program.

However, realization of these ideas required changes within the Department of Defense. Complex changes such as the restructuring of large-scale business procedures and the retraining of personnel were needed. At President Kennedy's assumption of office in 1961, a reshuffling of several important professorships at Harvard University, and the selection of Robert MacNamara as Secretary of Defense, created the requisite favorable circumstances for this task. In 1961, and the years following, the necessary reorganization was completed. As a result, a regularly functioning system of armaments planning and financing was created called Planning, Programming, Budgeting System (PPBS), and the new military budget was prepared on the basis of the applied methodology of systems analysis.[13]

In 1964, the procedures for using systems analysis in the Department of Defense were standardized to such an extent that some of them were published in the form of handbooks and instructions. The same kind of standardization was carried out within the National Aeronautics and Space Administration and many types of manuals and instructions were distributed to contractor firms.[14] All of these measures had a great effect upon armaments planning, and the allocation of raw materials and manufactured products. In August 1965, President Johnson ordered the adoption of the principles of programmed management and systems analysis in the majority of Federal institutions, including the Department of State.

Assimilation of PPBS did not take place without difficulty: Reorganization of the Department of Defense on a functional basis was not completely successful, and it became necessary to rearrange the new organizational functions among old organizational subdivisions. The adoption of new methods met with opposition before PPBS was fully operational. In 1968, the conflicts reached the point of senatorial inquiry: accusations against PPBS were investigated as well as the arguments against systems analysis, with its demands for a quantitative comparison of alternatives.

13. Hitch (1955) pp. 45, 83, 93.
14. F. Gedes (1964).

The reasons for the inquiry had to do with conflicts of interest of military-business corporations versus the interests of central military management. There were also difficulties with the assimilation of the methodology . . ., difficulties which were exacerbated by swift progress in the use of new techniques. There were other difficulties with the normative character of the methodology which must have played their role, also.

Somewhat behind the pace set by the military establishment, systems analysis came to be utilized in American industry and in other areas of society. Although from the earliest time Hitch considered systems analysis a military necessity, he also believed it was a means of supporting objectivity in industrial management (see Hitch, 1955). History has proven him correct: It has turned out that the breadth and diversity of the applications of systems analysis in civilian life are fully comparable to the scope of applications in the military area.

The use of systems analysis for problem solving in civilian life emerged in the mid-1950s. One of the first papers was the contribution of McKean (1968) on the development of governmental policy in the area of water resources. In 1959, several papers were completed on the use of electronic data processing in urban planning.[15] Since 1960, many investigations have been conducted on supersonic transport aviation as a total system.[16]

A considerable number of systems-oriented projects have been completed by the TEMPO division of General Electric using the methodology of systems analysis.[17] This organization has, for example, applied systems analysis to the problem of long term corporate strategy and policy making. Included in these efforts was an analysis of future alternative marketing strategies, and the effectiveness of marketing expenditures. Later, TEMPO conducted research on the development of atomic merchant marine through the year 1985, taking into account five alternatives with considerations of political and social environment. TEMPO also conducted estimates on commercial systems of communications using artificial satellites; it investigated the capability of North America in water resources and sources of energy, the development of the gold-mining industry in the United States, and the problems of urban development. In 1966, General

15. Optner, 1959. See also Zwick, 1962 and 1963.
16. ICAO document, 1962.
17. Information about TEMPO, The RAND Corporation and Stanford Research Institute may be found in Jantsch, 1967.

Electric conducted an analysis of the strategy of the construction of atomic power stations operated on plutonium reactors in Europe.

Even in the early period, The RAND Corporation used systems analysis to study extraterrestrial transportation in 1990. Systems Development Corporation, a company originally organized by RAND, studied problems which would lead to advancement in educational planning, medicine, urban systems and law enforcement soon after it was organized. Stanford Research Institute (SRI) applied systems analysis to problems of interaction between science, technology and society at an early date. In 1967, a detailed systems analysis was conducted on the supply of humankind with food. SRI has also applied systems analysis to the problems of industrialization of developing nations, and to the design of safeguards for developing nations in case their economic interests are threatened by expanding technology. Reconstruction of organizations along the lines suggested by systems analysis has been performed in hospitals, in sales departments of commercial companies, in electronics, manufacturing firms, in engineering, in chemical-pharmaceutical companies and in construction companies (see S. Young, 1966).

Following the United States, and under the influence of work performed in the USA, other nations have made use of systems analysis, first and foremost in military matters, but also in governmental institutions. Some of the countries using systems analysis include England, France, the Free Republic of Germany, Japan and others. In England, there is a long legacy of systems thinking (originally called *operations research*, or *operations analysis*) which started during the Second World War. The teaching of courses in systems analysis is the rule in many educational institutions in the USA and in other countries. For example, systems analysis is so well established, it has become a part of a program for the preparation of students for managerial posts at the University of Khartoum (see Hanica, 1968).

Systems analysis has also been applied to areas in sociology, politics and studies of ideology, each having its own specific problems. Improvement of the techniques for dealing with political reversals and maintaining control over populations . . . in particular with the aid of mass communications . . . may also become one consequence of the methodology of systems analysis (see Easton, 1965).

Systems analysis is the methodology by which data processing applications are organized for computer processing. In a limited period of time, so many different computer-driven systems have been created that it seems to be more correct to speak of an 'explosion' in this area. A 1964 list contains 133 systems projects in a total of 98 aerospace companies.[18] A 1965 list contains 65 systems projects in only a single institution, the Department of the Navy. Today, the literature contains tens of thousands of systems in all facets of society: the scope of these systems may be national, international or even global in character.

The creation and use of automated navigation systems has been influenced by the use of systems analysis. Semi-Automatic Ground Environment, a computer-controlled command system which pinpoints the location of military and commercial airplanes over continental North America is also a product of the new systems-oriented thinking.

The question of the relationship between the development of systems analysis and the creation of automated guidance systems is rather complicated. It would hardly be correct to assert that the development of systems analysis served as a precondition for the development of computer-driven navigation systems. In parallel with both developments, companies generated programs to automated clerical operations, accounting and planning work, as well as to improve organizational systems and procedures. The evidence in this area points to a synergistic condition where the problem and the methodology catalyzed each other.

In part, the development of technically advanced hardware and systems projects were influenced by application of operations research and econometric methods. As computers and programming techniques increased in power, there was a growing awareness among its users that automation and systems analysis could have a profound impact upon the solution of society's most acute problems. However, none of these new tools and techniques created dramatic improvements, unless the organization underwent changes to accompany the heightened effectiveness of its systems. At the present time, the place of computers, software, automated systems and systems analysis is becoming increasingly well defined. The most advanced systems are the man-machine integrated systems, which are oriented toward specific problems.

However, even today, the methods of formulating and applying systems analysis are far from being established. Several different points of view exist, even with respect to such a general issue as

18. R. Frambes, 1964.

the procedures for problem-solving. However, the general direction and developmental tendency of systems analysis are well established, and this is what serves as the basis for literature, lectures and real-life applications.

Even though periodicals heralded the development of systems analysis from the moment of their appearance, monographic and scientific literature only began to emerge fifteen to twenty years later (classified or internal expositions were issued earlier). In this way, current literature on systems analysis is based upon practical experience with its application.

An early publication of the applied and methodological character occurred in the early 1950s. In 1952, The RAND Corporation prepared a paper on Mathematical Models. This article was published by the University of Pittsburgh. The RAND Corporation later attempted to apply the systems methodology to the problems of organizations (in 1959). Members of RAND who made significant contributions to systems literature in this early period included Chapman, Newell, Byla and Weiner (1953, 1955, 1957). Newell later (1971) became co-author with Simon on many well known works on the theory of problem solving, heuristics and on an automated generalized problem-solver.

The importance of the earliest period can hardly be overestimated. One of the most significant and noteworthy books which appeared in 1948 was Norbert Weiner's. In 1951, Morse and Kimball's book, and Bertalanffy's article appeared; in 1951, his book was published.

One of the first books on systems analysis was published by The RAND Corporation in 1956. One of the most important books on systems design came out in 1957, by Goode and Machol. One of the first books on problem-solving appeared in 1955 by E. Hadnel. Kenneth Boulding's article, expanding on Bertalanffy's ideas, came out in 1956.

Up to the end of the 1950s, the periodical literature turned its attention to the classification of the differences between systems analysis and operations research (Sengupta and Ackoff, 1965), between systems analysis and systems design (Wohlsteffer, 1958), and between decision theory and operations research (Arrow, 1957). Issues such as the application of scientific methodology to the inexact areas of management, as well as the human solution of human problems and organization were widely discussed (see Weinwurm, 1952). In 1959, the well known book of Stafford

Beer was published, partly inspired by the ideas of general system theory.

Business and government-oriented books on systems analysis began to appear in 1958. It was in this year that McKean's book, containing an analysis of US water resources was published. In 1960, the fundamental book of Hitch and McKean on the application of the methods of quantitative comparison of alternatives in the solution of armament problems was made available.

The same year saw the publication of Optner's first book. In 1962, a fundamental book by Arthur D. Hall was published, as well as a book by D. O. Ellis and P. D. Ludwig. The periodical, *Operations Research*, also published a number of papers in 1962 on the principal questions of systems analysis.

In 1964, additional important papers by E. S. Quade were published for internal use by RAND. In 1965, a number of books of different character and orientation appeared: a second book by Optner [which is translated into Russian language]; the book by McMillan and Gonzales, containing a formal exposition of the subject, and the lectures of Hitch on the adoption of programmed management and systems analysis in the Department of Defense. Young's book (1966) was the first to devote itself to the systems analysis of organization as such. In the spring of 1968, the book by Kleeland and King was published, as well as a second edition of Optner's first work.

Contemporaneously with the books on systems analysis, books on systems design, which contained points of contact or of intersecting areas, were published. One of the first to attempt to formulize the theory of problem solving into the general theory of systems, was the undertaking of M. Mesarović in 1960. In subsequent years, other important books by Quade, Mesarović and Bellman appeared. The 1960s saw a number of published books on problem solving such as that of Kleinmentz (1966) as well as many on specialized aspects of the problem-solving process (e.g. Rightman, *Cognition and Thought*, or Miller, Pribram and Galauter, *Plans and Structures of Behavior*). The influence of systems concepts can be clearly traced in the growing wave of literature on the theory and problems of organization (e.g. *Approaches to Organizational Design*, by J. D. Thompson).

Considerations affecting future development

In those same years when systems analysis emerged and was developing, both in the USA and other countries, a number of new disciplines with unusual and intriguing names appeared: Operations research, heuristics, decision theory, system design and general theory of systems, among them. The introduction and newness of all these disciplines in such a short period of time gave rise to questions as to their boundaries, mutual relationships and positions within the scientific complex. Since the resolution of these questions would affect the interests of their champions, discussion took on the character of debate. This discussion was quite useful, and it continues to this day. There is a justification for these debates, and it is clear that systems analysis has already had an influence, and given a certain determinance to the discussion.

One of the methods of determining the relevant positions of each of the above areas of knowledge and activity consists in clarification of the differences and/or similarities in their functional assignments. This in turn entails a differentiation between functions and the method used for their realization.

The 'solution' of problems has 'non-solution' as its contrary, in the fulfillment of routine operations. Their synthesis may possibly encompass all areas of human activity. Problem solving is the basis for both the function of conservation and function of development. By its very nature, the problem-solving function is akin to the highest levels of representation of human activity. Hence, it is not surprising that the function itself, and the methodology subservient to it, are both in the highest degree integrated, and include within themselves all particular functions necessary for their realization. Systems analysis as a problem-solving methodology claims to fulfill the role of a substructure, integrating all necessary methods and activities for the solution of a problem. It is by this precisely that its relationship to such areas as the control of operations, the theory of statistical decisions, the theory of organization and the like are determined.

It is evident that there will be as many particular functions (and classes of methods used for their regulation) as are established by the *a priori* structure of the system methodology. Although it is not asserted that in actuality there are a precise number of functions, it is postulated that there are a specific number, given the axioms and the state of development of the system methodology. The determination of an exact list of particular functions for a particular system of axioms is a subject of specialized research whose necessity and importance cannot be overestimated.

However, a list having a provisional character may be set forth. Some particular functions include:

1. Symptom identifications.
2. Determination of the reality of the problem.
3. Definition of objectives.
4. Exposition of the system structure and its defective elements.
5. Discovery of alternatives.
6. Determination of the structure of alternatives.
7. Evaluation of alternatives.
8. Construction of solutions.
9. Acceptance of a solution by executive management.
10. Setting in motion the process of solution implementation.
11. Management of the solution implementation process.
12. Evaluation of the installed solution and its consequences.

Correspondingly, for the execution of the above functions, the following techniques may be used:

1. Methods of the theory of search and discovery.
2. Methods of the theory of symbol recognition.
3. Statistical methods.
4. Models of operations research and related models (queuing, hoarding, game theory, conservation and reconstruction, growth, etc.).
5. Models of behavior (homeostatic, dynamic, self-organizing, etc.).
6. Models of classification theory and ordering.
7. Marginal analysis.
8. Methods of synthesis of complex dynamic systems.
9. Theory of potential attainability.
10. Models of the theory of self-regulation.
11. Forecasting methods.
12. Methods of engineering psychology and related disciplines.
13. Methods and models of different areas of organization theory, social psychology and sociology.

Historically, the matter stood this way: in the absence of a problem-solving methodology, and in the absence of well developed methods for understanding the particular function of the process, specialists attracted by the problem would simply fall back upon their specialized areas of knowledge . . . mathematics, physics, economics, sociology etc. which became for them, the point of departure. Application of specialized knowledge, in some cases, led to success in understanding a particular function in the solution of a particular problem, and at the same time created a basis for the development of a method for the more precise understanding of this function.

However, the obvious inadequacy of any one particular method, no matter how powerful, to the solution of a problem made it necessary to surround it with auxiliary methods, in general not derivable from the method itself, either by recommendation, rules or precautions etc.: this was clearly visible in attempts at generalization within the framework of operations research.

The generalization of a particular method, or a group of methods, as is quite natural, soon claimed the position of being a problem solving methodology. This served as impetus to the development of a more comprehensive problem solving methodology, as well as to the fruitless discussions concerning the boundary issue, and the relative importance and significance of each contributing discipline. At the present time, when the methodology of systems analysis has determined, at least in part, the structure of problem-solving functions, such discussions lose their point (although, of course, the more exact definition of the limitations of particular disciplines has to continue); attention is concentrated on the identification of methods and problem-solving functions, as well as on the development of methods for those functions which are not satisfactorily understood.

Thus, in evaluating the interrelationships between systems analysis and the disciplines akin to it, and also in a determination of their positions within the other sciences, one should abstain from a discussion of the content of these disciplines on the basis of their names; one should abstain from attempts to deduce from their claims certain positions reflecting group interests; and also, one should abstain from giving in to the hypnosis of broad movements which several of these disciplines have attained. One should separate the actual achievements of each discipline into two parts: its contribution to the problem-solving methodology; and its contribution to the development of methods for the understanding of particular functions of problem solving. Such a regrouping of material would leave room for discussion of a purely semantic nature.

Another group of relationships among the disciplines considered may be clarified if account is taken of the specialization of the system methodology itself. Such a specialization, in diverse forms and in many directions, occurs almost from the very beginning of systems analysis.

One large, fairly autonomous area, that of problem solving with the aid of technical systems, has solidified into systems engineering; another one, still in the process of formation, is business system design. Specialization occurs also by way of subspecialities of systems analysis: in particular, the methodology of quantitative comparison of alternatives; the development of alternatives with respect to problems being solved (an area represented by such systems as PATTERN-QUEST (see Justice, 1964 and Cetron, 1967)); the selection of objectives, and others. Various areas of application, such as transportation, urban development, political and social areas . . . all these also give rise to their own specialized forms of problem solving.

The tendency to alter the relationship between separate scientific disciplines, so characteristic of the systems methodology, will continue to operate in the future. This compels one to recall the opinion given by Kenneth Boulding who predicted the appearance of a new 'system' of classifications of the sciences.

One of the most important characteristics of any methodology is its limitation. This is especially true of such a powerful methodology as systems analysis. An understanding of the currently existing limitations of systems analysis is just as necessary as an understanding of its capabilities. The limitations of every kind of tool used by human beings are determined by the area of application of the function performed, and by the degree of perfection inherent in the methods used.

From the description of sample applications given earlier, it is evident that the area of application of systems methodology is very broad. In principle, this methodology may be utilized in any area. At the present time, systems analysis is being used in military and in business development areas. In other areas, applications at this time have the character of isolated attempts. After exclusion of the applications area as a parameter in evaluation, the remaining parameters available for evaluation are functions and methods.

Concerning functional nomenclature, there are varying opinions in the American literature. Some, such as S. L. Optner, are inclined to suppose that systems analysis must include within itself all of the particular functions necessary to the solution of a problem. Here, even such marginal functions as diagnostics and development of alternatives (by means of a relevant organization of scientific research) are also included in the list of functions in systems analysis. Others, such as P. Clark, separate diagnostics, definition of aims and development of alternatives from systems analysis proper, and suppose that they each form autonomous scientific and organizational areas.

These differences are not sufficiently important to warrant contrasting them explicitly in an evaluation of systems analysis. The viewpoint 'everything that is necessary for the solution of the problem' is the more general one. In any case, in the question of boundaries, it appears to be more logical to consider the general situation in order that the more important limitations may be discovered.

If one supposes that problem solving includes all functions . . . from the impetus to the discovery of a problem, to the evaluation of the outcome of the solution adopted (decision taken) . . . one may then pose the question: do any other functions, performed by human beings, exist apart from the functions determined in this way for the solution of the problem? The only remaining area which lies outside the limits of this definition is the tremendously broad area of routine operations which actually becomes the core of problems. If this is so, then the role of systems analysis, or more accurately the role which lies behind systems analysis, and the role with which it will have in the future, is difficult to over-estimate.

The correctness and completeness of the nomenclature of particular functions, as well as of their content (i.e. the correctness and completeness of the structure of the subprocesses of problem-solving), may be determined by analysis of the adequacy of the nomenclature and content of the concepts used in systems analysis. They may also be determined as a result of evaluating the practical applications of systems analysis. One may expect that definite corrections will emerge as a result of this technique; however, an increase in the scope of a normative theory is, as yet, regarded as undesirable.

With respect to the excellence of methods utilized in understanding [the content] of functions: without going into detailed analysis of this question, it may be said that systems analysis, in spite of convincing successes, must yet undergo considerable development; in other words, it is still in its infancy.

Its main inadequacies are consequences of a pragmatic understanding of the term 'development': the idea of an objective development is replaced by the idea of a 'desired aim'. It is clear that a 'desired aim' may easily lead to false problems. Since problems are not defined as the objective contradictions of development, it follows that contradictions may increase in number in spite of the solution of problems. The pragmatic structure of methodology which ignores the objective character of the laws of dialectics, may seriously restrict the scope of the methodology.

A number of particular functions are still without an adequate tool for their understanding. These include:

1. Diagnostics of the immediate state of the system.
2. Diagnostics of organization.
3. Determination of the defective elements of the existing state of the system.
4. Methods of defining a nomenclature for alternatives.
5. Methods of delineating the tactics and strategy of the solution of the problem.
6. Identification of human factors for purposes of problem solving.
7. Defining preassigned types of behavior for organizations.
8. Evaluation of the consequences of the solution of a problem.

Intensive research is being carried on in many of these areas, especially in the areas of organization and management.

The structure of particular functions and the corresponding methods to deal with them must change when the concept of 'totality' is operationally introduced; in this case, it would provide a more accurate description of the relationships between the objects, the system, the methods and the functions; the relationship between the structure and the properties of the [system] elements will then also be more accurately reflected in the methodology.

A special area is occupied by the techniques of selecting methods for implementing the problem-solving function. The greater the scope and complexity of the problem, the greater the effect of constraints (boundaries) of the problem-solving methodology. The elimination of constraints will give rise to many new disciplines and to considerable development in many existing areas. What has to be developed is a theory of relatively specialized totalities (systems), their appearance, growth and development, norms and pathologies, qualitative transformations and degradations. To this end, one will need a theory of a system environment, a theory of a hierarchically organized systems environment, and many others.

The role of several normative theories are, at the present time, much exaggerated; one of the tasks of future development, without a doubt, consists of placing limits on their utilization consistent with the exact role which they are capable of fulfilling.

Since the structure of systems methodology is quite complicated, and has a tendency to become more so, its correspondence with reality may be established by a correct utilization of more or less powerful formulizations. Generally speaking, an evaluation of the effectiveness of concrete forms of systems methodology constitutes a serious problem. For this and other reasons, it is evident that the initial step must be taken in the direction of the development of various kinds of formal theories of systems and problem solving.

References

ARROW, K. J. (1957), 'Decision theory and operations research', *Ops. Res*, vol. 5, no. 5.

BEER, S. (1959), *Cybernetics and Management*, English Universities Press.

BERTALANFFY, L. VON (1950), 'The theory of open systems in physics and biology', *Science*, vol. 3.

BERTALANFFY, L. VON (1951), *General System Theory*, Braziller.

BOULDING, K. (1956), 'General systems theory – the skeleton of science', *Management Science*, vol. 2, no. 3.

CETRON, M. J. (1967), *Quest*, status Report, IEEE transactions on EM-14, no. 1.

CLELAND, D., and KING, W. R. (1968), *Systems Analysis and Project Management*, McGraw-Hill.

EASTON, D. (1965), *A Systems Analysis of Political Life*, J. Wiley.

ELLIS, D. O., and LUDWIG, F. J. (1962), *Systems Philosophy*, Prentice-Hall.

GEDES, F. (1964), 'Customer-closed loopholes in programmed management', *Aerospace Management*, vol. 7, no. 4, pp. 50–53.

GOODE, H. H., and MACHOL, R. E. (1957), *System Engineering*, McGraw-Hill.

HADNEL, E. (1955), *The Art of Problem Solving*, Harper and Row.

HALL, A. D. (1962), *Methodology for Systems Engineering*, Van Nostrand.

HANICA, F. P. (1968), *Management, Education and Management Science*, Operations Research Quarterly, Special Conference Issue.

HITCH, C. J. (1955), *An Appreciation of Systems Analysis*, The RAND Corporation.

ILAO (1962) *The Technical, Economic and Social Consequences of the Introduction into Commercial Service of Supersonic Aircraft*, Document no. 8087 c/925.

JANTSCH, E. (1967), *Technological Forecasting in Perspective*, OECD.

JUSTICE, A. (1964), 'Pattern', joint National Meeting of ORSA and IMS.

KAHN, H., and MANN, I. (1959), *Techniques of Systems Analysis*, The RAND Corporation.

MCKEAN, R. N. (1958), *Efficiency in Government Through Systems Analysis with Emphasis on Water Resources Development*, J. Wiley.

MCMILLAN, G., and GONZALES, R. (1965), *Systems Analysis*, Irwin.

MESAROVIC, M. D. (1960), 'General systems theory', notes from class lectures delivered at the Case Institute.

MORSE, P., and KIMBALL, G. (1951), *Methods of Operations Research*, Jointly published by The Technology Press of MIT and J. Wiley.

NEWELL, A., and SIMON, H. (1971), *Human Problem Solving*, Prentice-Hall.

OPTNER, S. L. (1959), *The Feasibility of Electronic Data Processing in City Planning*, City of Los Angeles.

OPTNER, S. L. (1960), *Systems Analysis for Business Management*, Prentice-Hall.

OPTNER, S. L. (1965), *Systems Analysis for Business and Industrial Problem Solving*, Prentice-Hall.

POLYA, G. (1957), *How to Solve It*, Doubleday & Kean.

RAND CORPORATION (1952), 'On the use and limitations of mathematical models: the theory of games and systems analysis', *Management Science*, vol. 5, No. 3.

SENGUPTA, S. S., and ACKOFF, R. L. (1965), *Theory of Systems and Operations Research*, Transactions of IEEE on SSC, Vol. 1, No. 1.

WEINWURM, E. H. (1952), 'Limitations of the scientific method in management science', *Management Science*, vol. 3, no. 3.

WOHLSTEFFER, A. J. (1958), *Systems Analysis versus Systems Design*, The RAND Corporation.

YOUNG, S. (1966), *Management: A Systems Analysis*, Scott-Forsman & Co.

ZWICK, C. J. (1962), *Models of Urban Change: Their Role in Urban Transportation Research*, The RAND Corporation.

ZWICK, C. J. (1963), *Systems Analysis and Urban Planning*, The RAND Corporation.

Part Two
Applications and Continuing Development

A review of on-line systems exposes the reader to some successes in use of computers and systems analysis in space travel, control of airline reservations, military command and control, planning research and development and problem-solving techniques. The theory and practice of systems analysis have produced some solid accomplishments in their first decade of use.

In this selection of writings, there was an apparent shift of interest away from the 'theory', however, to the practice of systems. Discussion of the value of information, later to be viewed as a commodity for sale, and a preoccupation with the organizational role of the system replaced the discussions of methodology. Hardware and software have been integrated with men and facilities as systems analysts enter the second decade of applications.

The management information system concept explained by Ream and Kanter in hardware-oriented articles signals the development of a new kind of systems analysis product. Similar in some respect to the command and control concept stated in Reading 7, this is a new view of how the system should be organized to optimize costs and effectiveness. Reading 9 illustrates the level of detail of one step in the design of the management information system, while Jaffe's Reading 13 describes another. Kanter's view tends to deflate any obeisance to the concept itself, and strives to identify the conditions under which the control objective of the management information system may be realized. Thome and Willard explore the systems approach which combines concepts of the management information system and systems analysis, in their elaborate description of a planning procedure for a group of related space exploration projects.

9 Adrian M. McDonough

Information Management

Excerpt from A. M. McDonough, *Information Economics and Management Systems*, McGraw-Hill, 1963, pp. 131–44.

Information is the measure of the value [worth] of a message to a decision maker in a specific situation (Proposition 4)

In practice, life for a manager is a flood of messages varying in scope and significance. As a 'message centre', his effectiveness depends on how he arranges for messages to flow to him and how he filters out the more significant content of the messages received. The manager in each position at each level is the *only* instrument that can produce either informal or formal measures of values. Others can influence the factors that go into a given manager's measures, but the measures can operate only in his mind and can reach the impact stage only through his decisions and resultant actions. This idea is paramount in any information-value approach to management systems.

Prerequisites to placing values on information

Information is defined as the measure of the net value obtained from the process of matching the elements of a present problem with appropriate elements of data. Proposition 4 says the same thing, but in this reading it is tied to the environment of the decision maker. Considering the environment in which a message is received brings in the following prerequisites to placing values on a given message:

1. A statement or label to identify the area(s) of value under consideration, e.g. health, security, profits, accuracy, speed. Only as we arrive at a level of specific area identification is it possible (meaningful) to develop useful measures of information value.

2. A statement of the range of potential opportunities or hazards involved in the specific area of value. This is the estimate referred to as the 'outcome stream of payoff', or the 'reflected total value of a study'.

3. A statement of the stock of knowledge available that can contribute directly to achievement of the reflected total value of the study. This stock was described as the 'Initial study value', or the 'study value at a specific time'.

4. A statement of just how much effort we may have to expend to achieve a certain position in the range of potential opportunities or hazards. This estimate can then become the framework within which a particular message is evaluated. The question can be asked: Considering how far we have to go, how much does this message help us to get there? This statement of 'distance to go' is a useful benchmark whether we receive the message because we raised a question or because a message arrives with no special effort on our part.

In practice, these 'statements' are typically little more than intuitive hunches. I have phrased them quite formally here in order to demonstrate the kind of problem with which we are dealing. If we are to advance to an approach above the intuitive level, some sequence of steps such as the above must be followed. Putting these statements in writing for a given study will provide the first step in information evaluation. I recognized in my earlier comment that there are many messages and many problems for every organization position. The formal use of the above statements can be applied, of course, only to the more significant problem responsibilities. Let us consider some of the factors that can influence the content of the above statements and their relationship to significant problem responsibilities.

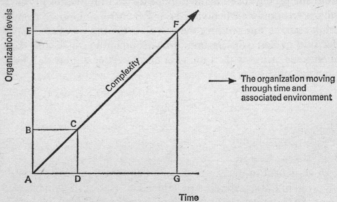

Figure 1 Problem complexity as a function of organization level and time dimension of decision

Figure 1 shows a generalized diagram of certain relationships regarding the factors influencing position responsibilities from a problem-content point of view. The overall diagram represents an organization moving through time and a changing environment. In each new time period new problems can arise to be assigned in the organization. Time is shown on the horizontal axis. At any particular moment in time, such as at A on the base line, different positions will be responsible for problems with different spans of time involved. Organization position levels are shown on the vertical scale. There is, normally, a tendency for higher-level positions to be responsible for problems with more factors or classes to consider, and also with more abstract factors or classes. Considering position responsibility from both a time span and an organization level makes it possible to bring these factors together on the line AF, labeled complexity. For example, area $AEFG$ represents a higher-level position at time A. As compared with a lower-level position $ABCD$, position $AEFG$ will have problems with:

1. Longer time spans with associated uncertainty.
2. More classes.
3. More abstract classes.
4. More criteria and weights needed.
5. Fewer criteria and weights available.

Proposition 4, to repeat, says that information is *the measure of the value of a message to a decision maker in a specific situation.* I am suggesting that organization level and time are significant factors to consider when viewing information economics from a total-organization point of view. The organization level implies the talents that will be used to seek solutions, and the time dimension carries with it the notion of problems presented by a changing environment. Perhaps more importantly, the organization levels can be considered for the specialization of problem-handling talents for each position at each level. It is this tie that allows us to bring in the above five factors of specific content of problems in a given position.

The statements regarding prerequisites to placing value on information and the organization level and time-dimension framework are proposed as general guides for placing value on management information. In the next few pages I consider organization levels and related complexity.

Value of information related to organization level

Assume with me a situation where a decision maker is under pressure to reach a decision; he gains by a good decision; he is harmed by postponing a decision or making a poor decision. If we could measure the pressure on the decision maker in this specific situation and compare it with pressures under an array of other decision-making situations this same individual might face, we might establish a scale of values for the array of problems. But what do we mean by pressures? Are these the emotional highs and lows of the individual, or can we establish a more concrete type of measure? Let us try a few possible benchmarks that may be helpful in analysing this measurement problem.

I am going to assume a scale of seven levels of authority and responsibility in the typical organization. Just for perspective, I am also going to weigh each of these levels by a hypothetical salary, e.g.

1. $100,000.
2. 60,000.
3. 30,000.
4. 20,000.
5. 12,000.
6. 8,000.
7. 5,000.

These dollars are one possible measure of the values of the total decisions related to problems made at each of the levels. They should represent at least this amount of benefit in money terms to the organization. A well-balanced organization presumably would get its values from these salaries by the careful placement of problems from its 'problem inventory' into these positions.

Value of information and level of problem complexity

Let us now consider the measurement problem from the point of view of the kinds of inputs possible at each of the organization levels. I shall picture an input as being a package that is made up of both a problem and the relevant classes, relations, and weights of data that can be used in the solution of the problem. I am viewing the problem-solution sequence as one of concurrent problem definition and solution. For our purposes here, let me call such an input a problem module, the implication being that we shall attempt to build jobs and overall organizations from such modules. With this possibility in mind, let us now consider a differentiating scale of classes of complexity for the kinds of materials with which the mind may work in the various levels of the organization. Chester Barnard (1940) has given us such a set of classifications. In the quotation that follows I have reversed the order of his classes so that I could align them with the position levels in the above list:

The nature of the material to which the mind is applied

The nature of the subject matter to which the mind is applied determines the kind of mental process which can or cannot be used. It seems to me this is apparent from a consideration of the various kinds of material. Without attempting a classification that is technical, precise or exhaustive, we may consider the material that we have to work with, of three classes:

1. Material of a speculative type
This consists of impressions and probabilities not susceptible of mathematical expression and purely contingent uncertainties, including the possibility or the probability of the existence of unknown factors and their possible effect. This type of material is an important and vital part of what the mind must be applied to in ordinary affairs, so much so that much of the skill required in the management of business and public affairs consists in avoiding so far as possible positions of uncertainty; but, in the nature of things, only a relative success can be attained. No matter how expressed, the processes involved in reaching a judgment here are subjective and nonlogical.

2. Material of hybrid character
This consists of data of poor quality or limited extent, propositions recognized as of doubtful validity or of tentative character, and qualitative facts which cannot be expressed numerically, requiring such adjectives as good or poor, bright or dim, orange or yellow, fine or coarse, stable or unstable etc. A large part of the mental work in business and government is certainly applied to material of this character. As its quality decreases, reasoning applied to it becomes more and more hypothetical and is quite speculative. The form of logical inference may be preserved but the premises become more and more mere verbal expressions without definable content and the reasoning mere rationalization, judgments and intuitions in the verbal form of thinking.

3. Material that consists of precise information
This consists of observations from which a conclusion may be drawn by scientific method, and propositions or facts previously established or widely accepted as true, to which formal logic can be applied. A great deal of such material is now involved in ordinary affairs, as is evident in legal work and in the great extension of cost accounting and statistical methods, and in other highly technical processes in business.

Much of the progress of recent years in many businesses has been achieved by developing methods of measurement that furnish material of this type. Of course, statute laws, the provisions of contracts, legal decisions and definite formulations of policy are also material of this kind.

We may generalize that the more speculative materials described by Barnard tend to be inputs at the top of the seven-level salary scale and the more precise materials are at the bottom of the scale, with the hybrid class in the middle range of jobs. These are only tendencies, for of course some of each class appears at each level. If we accept these generalizations, we are led to the primary difficulty of establishing simple measures of the values in the decision process, especially measures that could be consistent throughout the range of levels of positions. The benchmarks which will be followed, therefore, are selected to recognize the need for some set of scales of complexity for the total range in an organization. Such a scale has to be defined in some context. A basic context would be the effect of a particular message on an individual's performance, and particularly on his performance as it is evaluated by his superiors. The scale would go from worse off to better off, with a midpoint that indicates stability, i.e. no better off or no worse off because he has received a particular message. This scale can be useful only if it is established at each level up to the highest position in the organization.

Value of information in the context of the opportunities and hazards of the organization

The environment of an organization as it generates problems was described earlier as the prime reason for any work in management systems. Problems reflect from situations where opportunities may be obtained or where hazards can be prevented. From a given position at a moment in time, an organization seeks to take actions that will increase its position (opportunities) and at the same time to take actions that are of the nature of maintenance of position (hazards). Of course, this same reasoning holds for the action of the individual in the context of motivations and values. As the organization moves through its environment, individual participants identify their place in the scheme of things. They consider their organization position level and the time dimension of decisions for which they are held responsible. It is at such a time that the individual uses his management vocabulary to identify the materials (Barnard's classifications above) with which he will work and, at least intuitively, starts to assign priorities to his problems. In so far as his opportunities and hazards are synchronized with those of the overall organization, he weights his problems by their importance to the organization. When the individual recognizes a lack of consistency between his personal interests and those of the organization, he is faced with the forces of compromise. The scales of opportunity and hazards that he sets for himself will be balanced by the repercussions he can expect if he gets out of line with the best interests of the organization. In any case, if the individual is to make logical decisions and is to take logical action, he must be as explicit as possible regarding both his opportunities and hazards. Again he must select and express himself in the criteria peculiar to his situation.

Value of information related to problem definition, information gain, and decision-gap perspective

Under Proposition 1, I suggested that the position of a qualified individual be spelled out in three statements:

1. Statement of the problems.
2. Statement of the knowledge and data available to meet the problems.
3. Statement of unknowns or decision gaps. A decision gap was defined as the conceptual distance between the point where we stop further formal analysis, take our best view of what we see, and leap through or over what still remains unknown to a decision.

A desirable scale using these three statements would be one that recognizes the contribution of a particular message to:

1. The improvement in statement (definition) of the problems.
2. Increased information to offset the problems.
3. A recognition that it will be too difficult or impossible to get some of the desired information and that a decision gap exists. The contribution here is that the individual stops looking and faces up to his decision with the available information.

The information value model for management systems now has five benchmarks:

1. Levels of authority and responsibility with indicative salaries.
2. Time dimensions of decisions.

3. Levels of problem module complexity.

4. Opportunity-hazard scales.

5. Contribution to problem definition, information gain, and decision-gap perspective.

These benchmarks can be used as a basis for judging the net-value change that takes place because of a given message to a specific individual responsible for a particular problem module. The net value of the message is the sum of the values reflected from improvements in problem definition, information available, and decision-gap recognition. These values in turn are derived from the pressures on the individual because of the complexity of the problem, the importance of the problem on his opportunity-hazard scale, and the relative significance of the level of his job position.

The purpose of a system is to carry information to decision makers (Proposition 5)

Above the level of mechanical processing[1] there is no justification for handling data unless there is some sort of decision being made with regard to the value of the data. This is particularly true at the end of the stream of information flow, i.e. when a message is received and it influences the actions of the receiver. It is also true all along the line as the message is being put together for its end use. Each choice of message content is a decision, for each choice of content can influence the final message, and therefore the actions that may come about. Each elemental sequence in a system therefore reflects to and from the ultimate actions that are feasible as a result of decisions influenced by the system.

Thus, while this proposition states that the purpose of any *system* is to carry information to a decision maker, it also recognizes that the purpose of a *system's designer* is to achieve a balance between the decision mechanisms built into the system and the decisions that individuals make as a result of the system's output. This is one of the reasons why systems-design work can be so challenging. This is where progress is made. It is also the specific involvement where the competition between systems and individual becomes most apparent. Within the system, the systems designer

1. In mechanical processing we see that systems can have a built-in capacity for making choices, i.e. decisions. In operation, systems can receive information to make decisions previously made by personnel.

determines the kinds and ranges of choice. He also determines the type of information service that will be supplied through the system to users of the system. Of course, in both cases he is guided (or restrained) by others who have interests in the problem area.

Our interest is in the information support provided by a system to the decision maker. Note that I have shifted in the preceding sentence from a system as competition to a system as support or aid to the decision maker. The point to be made is that a given system can be negative and competitive if it jeopardizes some individual's market for his knowledge. The same system can be positive and reenforcing if it complements, and therefore extends, an individual's performance. What follows in this reading concentrates on the supporting aspect of information services provided by systems to decision makers.

Any system is a logical configuration of the significant elements in a selected problem area (Proposition 6)

A ———— system is a logical configuration of the significant elements in a selected problem area. The unique factor in this definition is the blank before the word 'system'. It reminds us that we must identify the particular field of knowledge in which we happen to be seeking improvement. Note that the definition is not limited to business systems. What I have said so far could be said about any field of knowledge, or knowledge in general. The systematic approach is a common denominator for advances in all fields of knowledge. For example, we can say that a legal system is a logical configuration of the elements of criminal law, maritime law, contract law, patent law etc. The medical researcher and the chemist, the lawyer, and even the musical composer have their systems. Each has his approach to classification and analysis; each has his symbols and means of measurement. Each seeks to improve productivity within his particular field of effort.

The business systems designer applies his talents to the array of management problems and their requirements for systematic approaches to solutions. In business today we hear a lot about models. We can have three-dimensional models, graphical models, mathematical models, even models using electrical energy to simulate various conditions. These are the structures developed by systems designers as they speculate, test, and apply the results of their efforts. I suggest that the above definition is particularly appropriate for the development stage of systems design. A merit of viewing systems from this conceptual point of view is that it is consistent with the primary characteristics appropriate to the definition of knowledge itself.

The above definition can be treated as a system, and as such it can be considered in its separated parts. These parts are the key words in the definition, i.e. 'logical,' 'configuration of elements', 'problem area', and the two terms 'significant' and 'selected'.

Logical. According to the dictionary, 'logical,' as applied to results, means that they are obtained through a process of formal reasoning. In business we know that many decisions are made by the individual's use of his intuition, so I have used the term logical as applied to results to mean 'obtained through a process of formal reasoning, plus the support of intuition.' Note that there can be illogical systems as well, but for the purpose of conceptualization, I use logical in the sense of benefiting the organization. Logical means mentally created. Processes of logic, of the scientific method, where we hypothesize, analyse, bring back together in synthesis, test, and apply, are derived from the ability of the systems designer to reason, compare, and thus decide.

Configuration of elements. For the term 'configuration' in the definition you may wish to substitute other words, for example, model, structure, chart, or image. This definition of system, in and of itself, is a form of configuration. Configuration implies pieces that have to be put together, or a set of relationships. These relationships are between and among significant elements. Normally, they are narrowed or abstracted from reality, because reality is too complicated to handle with all possible elements in the configuration. The concept of suboptimization, heard in operations-research circles, is relevant here. We cannot have the configuration unless we have the pieces, or the significant elements. The elements are of no use unless they are put together into the configuration.

One of the difficult questions is, do we build down from the configuration to the significant elements, or do we build up from the significant elements into the configuration? Notice that this is the process of classification. Much research is being done in this area. The information-retrieval work now being given major attention is based upon classification.

To classify the elements of a problem, symbols are required. Here we have an example of classification. If we are going to communicate, we are limited to words, pictures, or numbers.

Thus we find the significant elements in business systems appearing as words, in reports; pictures, such as charts; and various diagrams and numbers in the form of accounting statements or mathematical programs. These are the languages and the classes with which a problem is analysed into its component parts.

Problem area What is a problem? A problem, according to Webster, is 'a question posed for solution.' From the point of view of management systems, a problem is the recognition of an opportunity or a hazard. The opportunity or hazard becomes the justification for investing resources to seek a solution.

What is a problem area? By attaching the word 'area' to the word 'problem' we are recognizing that we can think in terms of the dimensions of a problem. For example: is it a broad problem? Is it a narrow problem? Further, we can think in terms of the specific content within the area of the selected problem. For example: Is it a finance problem? If so, what particular elements of finance are involved? This sounds much like the 'configuration of elements' described above – and it should. Though we start a study with a simple one- or two-word label for a problem, at the end of a study we have defined the problem in the detail of its elements and their interrelationships. 'Selected problem area' refers to our initial short label for the problem, whereas 'configuration of elements' considers what we are trying to achieve during the study. Thus the definition includes the before-and-after view of a problem.

Significant and selected. The terms 'significant' and 'selected' indicate alternative choices. Any time there are alternative choices, measurement is required, comparisons must be made, and selection must take place. Thus again we come to the question of required measurements in the appropriate values of the particular system. These values are not necessarily constant, so policing action is necessary when we find the values changing with changes in the problem area.

There are at least two ways to look at the term 'selected'. In the first instance, let us assume that all problems are classified. Our systems problem then is, which one do we work on? If we assume a fixed inventory of problems and we have a large enough budget to employ enough people, we can work on all the problems But normally we have limited resources and must be selective in picking our problem areas. Second, and more complicated, is the assumption that the problem universe has not yet been subdivided. The discussions on the 'total system' concept center around this point. We are still trying to conceive and spell out a useful total system of a business, and some progress is being made. Note that automation – a term applied to the shop, and later to the office – is now being applied to an integration of both shop and office.

The question of selected problem area has at least two parts: (1) What is the totality of the system, as just mentioned? (2) How do we subdivide or set limits to the scope of the subdivision? Note that there are various restraints operating here. Let me cite just three restraints on how big a problem we can take on.

First, in practice, each of the separate systems, covering separate problems, has usually been developed to the limits of comprehension, or conceptual ability, of those involved in its design and utilization. At some point things go out of focus and out of control. A manager who does not admit the open-endedness of complication will pay the price as his 'ulcer quotient' goes up. Though there are no natural boundaries to systems, there are administrative limitations.

Second, there are no 'instantaneous business systems'. The business system is almost meaningless when pictured as a moment in time. If we think of a system as a logical configuration, it implies a logical sequence of steps, and for these logical sequences of steps there are time intervals as inherent factors in the systems. These time intervals may be quite different for various problems in the same business.

A third restraint would be that each element covered by a system is, in the final analysis, a product of human judgment applied to the classification of activities. These classifications may be appropriate in one period and less appropriate in another, for example, the first week in the month versus the last week in the month. Thus we must either have alternative classifications or, of necessity, work with averages or optimum classifications. We can say that separate systems covering quite different problems can often justify quite different classifications. Any attempt to standardize is accomplished only at the price of further averaging, and this barrier often sets a limit in systems scope.

What I am implying here is that, as we try to increase a system's scope or take on larger problems, there are compounding complications pressing back on us from these restraints. Total systems are not good in and of themselves. I might refer back to the production environment where we had something like a total system years ago, when one very large motor would drive many machines through a series of belts. Then we found it advantageous to set up separate systems with individual motor drives on machines. It was only when we were able to conceptualize the overall relationships of the individual machines that we were able to move to integrated production systems, and even now this covers but a very small percentage of our physical productive capacity.

A significant area of systems design is the setting of boundaries in which systems can operate with some measure of self-policing. The setting of boundaries is a key part of any problem definition. Proposition 6 provides an operational definition of the term system. It recognizes the importance of the initial selection of the problem area as an integral part of systems design. It also provides reminders of the need to keep our criteria clearly in mind throughout a systems study.

A cycle of business systems

Figure 2 shows the six information management propositions in a sequential format. It is proposed here that this sequence of six steps can be used in the development of an information support program in any type of organization. The line that feeds back from step 6 to step 1 says that the overall collection of

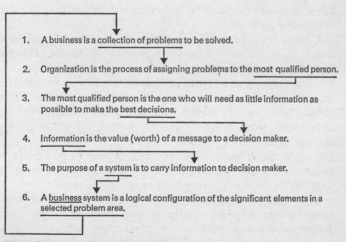

1. A business is a collection of problems to be solved.

2. Organization is the process of assigning problems to the most qualified person.

3. The most qualified person is the one who will need as little information as possible to make the best decisions.

4. Information is the value (worth) of a message to a decision maker.

5. The purpose of a system is to carry information to decision maker.

6. A business system is a logical configuration of the significant elements in a selected problem area.

Figure 2 A cycle of business systems

problems is made up by a selection of problem areas. What goes in at the top as the selected problems should be consistent with the ultimate output of systems produced. This line represents a reminder to see if, in fact, we are producing systems that are consistent with the priorities assigned to our key problems.

Reference

BARNARD, C. (1940), *The Functions of the Executive*, Harvard University Press.

10 Norman J. Ream

On-Line Management Information

Excerpt from N. J. Ream, 'On-line management information', *Datamation*,
Technical Publishing Co., 1964, pp. 27–30.

The past decade has seen the evolution of the use of electronic
computers and the evolution and recognition of formal management information systems. The planning and formalization of
these management information systems and the accompanying
necessity of integrating them on a company-wide basis is a massive and difficult undertaking. The development of integrated
information systems on a batch processing basis is difficult
enough, but when we consider an attempt to place such systems
on a true real-time basis, the problem becomes so large it staggers
one's imagination.

Paralleling the evolution of more formalized business systems
was the use of computers in process control, probably the first
major use of computers in the real-time business environment.
Examples are the introduction of numerically controlled machine
tools into manufacturing operations and the development of
computer-controlled chemical-process operations.

Closely associated with these latter developments was the
development of military command and control systems that
combined control of a process with the production of formally
structured information. An example here is the SAGE system.

A real-time management information system may be defined
as a system whose primary product is management information,
as opposed to systems, such as process control and command
and control, in which the production of such information is
treated as a byproduct. This field is new although there are
examples of limited applications of such systems. As in any new
area there are many problems to be solved.

Management and the information system within which they
must operate are inseparably interdependent. The accelerating
pace of technological advancement and the anticipated accompanying shorter product life-spans, together with the increasing

rapidity of changes in marketing climates, means that management reaction time to change in all areas must be speeded to the greatest degree possible. Consequently, we must recognize that management cannot continue to rely upon existing reporting cycles nor can they continue to place their dependence on historical information. Rather, management must have immediate access to the effect of changing conditions on their present management climate as well as a means of determining the effect of current conditions on planning strategy.

The problem facing management today is not what action should be taken to meet present conditions, for those actions should have been taken yesterday; rather the problem is what action must they take today to meet future conditions and to insure corporate survival.

My intent here is to point out some problems involved in implementing real-time management information systems and to suggest some solutions to these problems. Basically the problems fall into three categories: hardware, systems design and programming, and management use of these systems.

What is a real-time management information system?

A control system is a combination of a data processing system, a management information system, and a feedback system. If corrective action is taken while the process is still going on, we have a real-time control system. Further, there are three levels on which a real-time system may operate. First, if the system accepts input directly, processes it, compares it with predetermined parameters, and issues instructions to men and/or machines, we have a real-time control system. Second, the computer may inform affected parties of this information as it develops. This level may be termed real-time communications. Finally, suitable condensations of the information derived are prepared for higher levels of management. Here we have a real time management information system.

Generally then, we may define real time systems as those systems that keep pace with 'live' operations, accept data directly without manual conversion, process these data and establish relationships among data of disparate types. Further, they output data, on demand or as a result of programmed logic, to men and/or machines in a *timely* and *digestible* form.

For purposes of this paper, however, a line must be drawn

between command and control systems and real-time management information systems.

If a system exerts direct control over the physical environment from which it accepts data, I will classify it as a command and control system. Examples of these systems are the SAGE, BMEWS and Mercury systems.

We can define a real-time management information system as one which monitors the physical environment but exerts control only indirectly by the production of management type reports or displays. Examples here include existing airlines real-time reservation systems, various savings bank systems, and Lockheed's Automatic Data Acquisition (ADA) system. These systems may have primary functions other than management reports. However, they all use the data gathered by the computer to provide management with structured information. This latter type of system will be the one which this paper will discuss.

Why a real-time management information system?

In order to answer this question we must first define the term 'management information system.'

Managers need communications systems or reports. These reports may be considered under the general headings of Planning, Control, and Operating reports. In order to exercise control, management also needs specifications of objectives, criteria for evaluation of performance, decision rules for corrective action, and a feedback system to evaluate the effectiveness of corrective action.

Planning reports evaluate the position of the company in industry as compared to other comparable business entities. These reports include alternate courses of action available under a series of predetermined premises.

Control reports inform top management of operating performance as compared to predetermined performance standards.

Operating reports inform functional management of the current performance of operations within a given function. Normally these documents include a comparative analysis of current operations and operations for a previous period, as well as current performance as compared to predetermined detailed standards.

In addition to these reports, a real-time management information system can develop byproduct data to produce new criteria for performance evaluation, particularly of a statistical nature, at virtually no additional cost.

For instance, tighter control of materials and more efficient production scheduling are being realized through the Lockheed Shop Order Location system. Reductions in clerical and data origination costs are being attained in many systems through the use of real-time recording of payroll and labor distribution transactions. An increase in sales may be brought about by real-time inventory control which allows management to make better use of available inventory.

In addition, the use of a real-time system can make more profitable use of computer time. Using the classic batch-processing techniques of file updating and reporting, approximately 40 per cent of the computer time is spent in sorting. Additionally, about 20 per cent is spent on set-up time. The random updating of files eliminates most sorting and set-up time.

We are finding that in today's complex corporate world, data origination is rapidly becoming prohibitively expensive. The high cost of recording, accumulating and converting operating data to machine sensible language, combined with the fact that batch processing techniques cannot supply 'time current' information, have led to an increasing need for on-line, real-time systems.

Early real-time systems were usually quite large and even today most real-time systems being installed use large-scale computers as their central hardware. However, there have been real-time management information systems on a rather small scale.

As an example of one of these, an electric typewriter plant has installed a real-time quality control system centered around a small scale random access computer. The system has proved very profitable to the manufacturing operation and yet the total machine rental is less than $5000 a month.

Real-time system components

All real-time systems are composed of generally the same sub-systems (see Figure 1).

There is always a number of data acquisition or input devices, a communication system including any necessary interfaces, a central computer, and an input and output system within the computer center. Usually the system will also include remote inquiry and output devices.

Input sensor devices usually allow the use of a coded badge or card together with a punched card and variable keyboard informa-

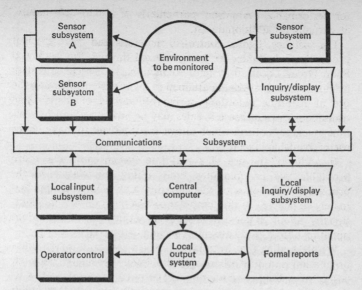

Figure 1 Organization of a typical real-time management
information system

tion. Needs in this area run to lower cost units of higher reliability
and more flexible operation.

Output devices are generally of one of two types. The printing
device such as the teletype page or strip printers and the flexo-
writer, or the newer CRT display devices. Requirements here
again are the classical ones of lower cost and greater flexibility.

The communications network invariably plays a large part in
the real-time system. Present networks commonly make use of
multipart cable, standard telephone lines, radio or TV channels,
or microwave links. Generally, capacity of these systems varies
with the cost.

The central computers in a real-time network may occur in
varying configurations and each of these configurations has its
inherent advantages and disadvantages. The choice of a particular
configuration depends in large part upon the system's requirements and in turn has a strong effect on the overall system performance.

Because the choice of a particular configuration is of such
importance to the design of a system and to its eventual success,
let us examine the seven basic hardware configurations.

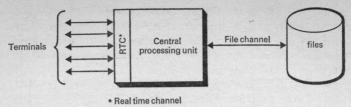

* Real time channel

Figure 2 Simplex system

The simplex system (Figure 2) provides no standby equipment. Among its advantages are: it is the least expensive from a hardware standpoint and, therefore, tends to be easier and cheaper to program as there are no complex routines for switching between two or more computers. Significant savings can be realized through the use of a simplex system, providing one is willing, or able, to lower system performance standards in the area of backup. The system must be off-line during maintenance, and recovery in the event of hardware unavailability must be manual. This, of course, requires that part of the savings in equipment and programming costs be reinvested in an extensive manual backup system.

If it is satisfactory, for instance, for the system to operate sixteen hours a day rather than twenty-four and if it requires, say, 90 per cent, rather than all of the transactions to be handled immediately, the simplex system may well be the best.

The next configuration is the simplex system with an input-output multiplexor (Figure 3). The multiplexor is a simple stored program computer which acts as an interface between the communication terminals and the computer. It may be a simple buffer or may be more sophisticated to the point of accessing the main computer only when access to the files is necessary. The advantages of this configuration include added modularity – changes can be made in the multiplexor to effect different scan rates, changes in priority etc. without disturbing the central processor

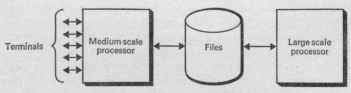

Figure 3 Simplex system with input/output multiplexor

programs. Also the memory allocations within the central processing unit are simpler because input-output and queue functions are controlled within the multiplexor. The disadvantages of this system are higher equipment cost, program interface considerations, and more complex reliability considerations. In addition, program testing begins to get complicated.

In this configuration (Figure 4) we have added a file multiplexor, which acts as an interface between the computer and the data files, to the previous system. Now the central computer is free of all specialized functions. The advantages and disadvantages are the same as for the previous system except that we have the added

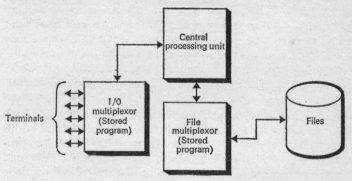

Figure 4 Duplex system with input/output and file multiplexors

advantage of removing file access considerations from the central computer and the added disadvantages of programming for still another machine and of further complicating hardware reliability considerations and testing procedures. Incidentally, a new possibility is opened here of a direct route between the I/O and file multiplexors, bypassing the central computer altogether except when processing is necessary.

Here (Figure 5) we have the first step in the configuration hierarchy that employs two computers. This system is well suited for any situation where a heavy load of internal computation exists. All housekeeping and scheduling functions are taken care of in the medium-scale master computer, leaving the slave, a powerful large-scale system, free to perform computations. Typically, the master receives data or information requests from the outside world and prepares all necessary tables, files, subroutines etc. It then sends the entire package to the slave which in

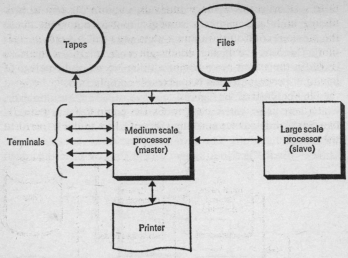

Figure 5 Master/slave system

turn performs the calculations and sends the resultant data back to the master for formatting and output. The advantages of this configuration are more computation capability per dollar, provided the medium-scale computer's cost can be justified, and automatic, one-direction back-up, i.e. the master can continue to receive, prepare, and batch input while the slave is unavailable. The disadvantages include high equipment cost and the complexity of the control programs.

The fifth configuration is called the shared-file system (Figure 6). It is quite similar to the master-slave configuration and in fact the master-slave complex could use the shared-file concept. Here again the medium-scale processor is the scheduler and controller and the large-scale machine is the computation device. There is added flexibility here because the medium scale computer can prepare jobs and place them on the file while the large system is busy.

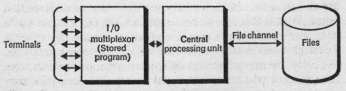

Figure 6 Shared-file system

194 Applications and Continuing Development

Both systems scan the file at intervals, the large system to pick up jobs and the medium size computer to get the answers. As in the previous configuration, this system need not operate in real-time. The advantages and disadvantages here are the same as those for the master-slave system with the added disadvantage of having to develop programs to accomplish data transfer between the file and both of the CPU's.

Here we have the duplex or dual configuration (Figure 7): two complete hardware systems, either of which is able to perform the total job.

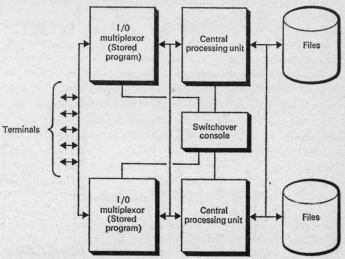

Figure 7 Duplex or dual system

In the duplex system the second machine is not on-line and must be switched over if the primary system fails or is taken off-line for any reason. The standby system, therefore, may be used for off-line batch processing, although it must have an interrupt built into its program monitor. In the dual concept both systems are on-line in parallel, performing the same functions except that output is generated by only one computer. Checking is carried on constantly, comparing the results obtained by the two systems. The dual system is generally considered the ultimate in reliability, but interestingly enough carries with it some fascinating problems, such as which machine is right when they are in disagreement? The dual system is, of course, more expensive than the duplex in terms of work accomplished, but also provides more reliability. Programming considerations are extremely important here as the complexity of control, monitor and switch-over programs may readily be seen. Most problems inherent in this type of system concern themselves with the transferring of information from one system to the other, the ability to update a system that has been off-line without interrupting the system that is functioning in real-time and the ability to preserve memory contents under an equipment failure. The general question here is 'How far does the user go in search of reliability?' Project Mercury is a full dual system while Sabre, the American Airlines reservation system, is a duplex configuration.

Finally, the multiprocessing system (Figure 8). Here we have two or more computer systems each doing more than one job. Standby in this system is on a 'degraded' basis. That is, when one system goes off the air another in the network may pick up the load. Because the computer picking up the load must also continue to perform its own tasks, a lengthened system response time

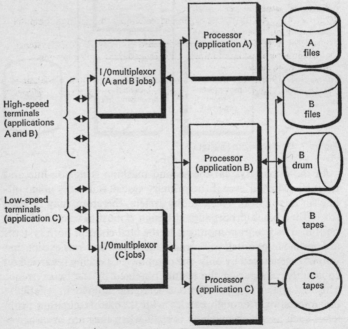

Figure 8 Multiprocessing system

usually results. Hence, the term, 'degraded service.' Depending on the configuration of the individual computers, any one may not be able to take over any other's task. These systems tend to be geographically separated. When one center goes off the air, another temporarily takes over its duties. SAGE has this type of standby ability. However, in addition, each SAGE site is duplexed as well as being a multiprocessing system. The advantage of the multiprocessing system is standby reliability at a lower cost than for a duplex system. The disadvantages include a lower quality of service when one system is down, the difficulty of performing reliability analysis on the system because of the complex interdependency of the units, and the fact that the planning, development and testing of programs and specially of control programs is extremely difficult.

References

HEAD, R. V. (1963), 'The programming gap in real time systems',
   Datamation, February.
HEAD, R. V. (1963), 'Real time systems configurations', IBM Systems
   Research Institute paper, April.
HOLDIMAN, T. A. (1962), 'Management techniques for real time computer
   programming', *Jour. ACM*, July.
HOSIER, W. A. (1961), 'Pitfalls and safeguards in real time digital systems
   with emphasis on programming', *IRE Transactions on Engineering
   Management*, June.
REAM, N. J. (1960), 'The need for compact management intelligence', in
   D. Malcolm and A. Rowe (eds.), *Management Control Systems*, J. Wiley.

11 Jerome Kanter

Integrated Management Information
and Control Systems

Excerpt from J. Kanter, *The Computer and the Executive*, Prentice-Hall, 1966, pp. 45–56.

The system

Integrated management information and *control systems* are fancy words which mean different things to different people. Let's approach the definition from the negative side by first indicating what such a system is not. It is not some weird, supposedly automatic solution to the problems of management, whereby management planning and decision making are delegated to a mysterious combination of robots and machines. This type of thinking is epitomized by the popularity of cartoons picturing the displacement of man by machine. A popular one shows a lone celebrator standing next to the computer with a party hat on his head and a drink in his hand mournfully exclaiming that 'Office Christmas parties were never like this.'

Neither is such a system a panacea – whereby management may merely convert its hitherto unsolved problems to some misunderstood series of binary numbers, sit back, watch the lights flash and receive an answer. Management does not become a passive participant, given the luxury to sit back and wait for the results. This concept of 'problems in – solutions out' can often result in 'problems in – greater problems out'. Research surveys have shown that a major reason why some companies are successful in employing the computer and others are not is *management involvement*. The management of the successful companies recognizes the potential of the computer and takes an active part in the determination of overall objectives and priorities rather than delegating these duties to subordinates.

Likewise, an integrated system is not just a happenstance – a thing that just came to be. Many systems just did 'come to be' but they are certainly not integrated systems. They are more like badly worn innertubes, patched and mended where conditions have necessitated expeditious and sometimes not well thought out solutions. In fact, in many frequently patched systems, a new patch will cause a blowout somewhere else. This type of crisis approach is a long way from an integrated system.

Let's bring into focus just what we are talking about. An *integrated management information and control system* may be defined as a management-oriented system conceived and designed by management as a single, total entity to control an entire organization. It does not evolve as a result of the development of many more or less independent applications. Individual applications are designed to meet the needs of a restricted area of the organization, but with the needs of the whole organization in mind. This results in a system welded together by data flows where redundancy in data storage and the transmission of useless information from one area to another are eliminated.

An *integrated management information system* begins with management. Management must determine the reports and analyses necessary to meet the requirements of their business. The paper work flowing from business operations acts as a mirror to reflect the real world of sales and production and finance that exists in running a business. The paper world must reflect this real world in the most realistic and meaningful manner.

After the needed outputs are determined, it is the responsibility of the data processing manager to produce these outputs in the most efficient way. In so doing, it is surprising to discover that almost all the information needed emanates from a rather small number of common source documents. It is also surprising to see the interconnection of systems which were thought to be entities in themselves but are really connected by a common flow of data. A prime example is a purchase requisition. This document initiates a long series of activities starting with the writing of a purchase order by the accounting department, and including the physical receipt of the goods, the quality control check, the paper work updating of inventory records, the establishment of accounts payable and eventual vendor payment. Thus the purchase requisition affects the accounting, financial, quality control, receiving, purchasing and inventory control functions. An integrated systems approach recognizes rather than ignores these interrelationships. The system may begin with the automation of a specific function but only after the total picture is studied. This permits the later addition of subsystems with minimum effort and duplication.

Let's now build an integrated system using the inventory control system previously described as one of the subsystems. It was made clear that sales forecasting is an important cornerstone of an inventory control system. Our integrated system will begin with the sales activity or sales subsystem. It has been often stated that 'nothing begins until something is sold.' This is certainly true and we will see that basic sales data is a primary information source from which flow a host of useful management reports and analyses. Each system will now be called a subsystem since it will be but one part of a single integrated management information system. Figure 1 is the starting point of the integrated system.

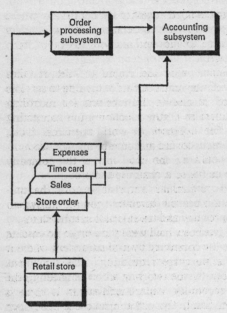

Figure 1 The order processing and accounting subsystems

It is obviously important to satisfy customer demand since this is the principle reason for being in business. Management should continually strive to give the customer the best product at the best price, and at the same time to realize a profit in line with the product being offered. The basic cycle starts here because what goes out of the store must be replaced. As the shelves in the store are depleted, certain store operations are put into play which result in the reorder of goods to replenish the shelves. The store will reorder goods on the basis of turnover and the lead time required to replenish the item. The store order is one of the basic source documents which initiates a long series of activities throughout many departments. This document enters the first of the subsystems – the order processing subsystem. This subsystem will screen the orders for errors in omission as well as commission and ensure that the order is valid before it proceeds further into other subsystems. The connecting arrow to the accounting subsystem indicates that the order is the basis for setting up accounts receivable and eventually reconciling the cash payment for the goods.

Another source document is the record of sales or the movement of items from the store. This source document will eventually produce meaningful sales statistics and analyses as output from the accounting subsystem.

In addition to the store order and record of sales, it takes people to run a store, and they must be paid according to services rendered. A time card represents the medium for recording hours worked; it is shown in Figure 1 entering the accounting subsystem. Likewise, the expenses of store operation (heat, light and maintenance) are recorded and entered into the accounting subsystem. This would be the case only if the company controlled its own stores (as in a chain operation).

Now we enter into more familiar terrain as we build the integrated system. Figure 2 adds the inventory control subsystem as well as the materials control and transportation subsystems.

After the store order has been validated by the order processing subsystem, it moves to the inventory control subsystem where it must be screened against the current inventory in the warehouse. This process is indicated by the two-way arrow connecting the order processing and inventory control subsystems. If there is sufficient inventory on hand, the order processing subsystem produces a store order picking document for the warehouse and an order invoice for the retail store. As employee labor and operating expenses were recorded at the store level so they must be recorded at the warehouse level. These can be seen entering the accounting system of Figure 2. The blank card will be explained in Figure 3.

The materials control and transportation subsystems are excellent examples of how an integrated system operates. For example, the basic information on orders and movement from the warehouse can be used to more efficiently lay out warehouse space. Likewise, the information on each order such as the cubic content and weight of items comprising the order can be used to load and route trucks. These valuable by-products can often be real payoff applications. Thus the same basic raw data is processed in a little different way to effectively help in the handling of another functional activity.

Figure 3 adds another subsystem to the overall integrated system – the purchasing subsystem. This subsystem is dependent upon the strength of the inventory control subsystem. A strong inventory control system provides for automatic ordering based on considerations such as customer service, economic order quantity, and lead time. Orders were screened against inventory records in Figure 2. When the inventory of particular items reaches a predetermined reorder point, the inventory control subsystem directs the purchasing system to write a purchase order, which can be

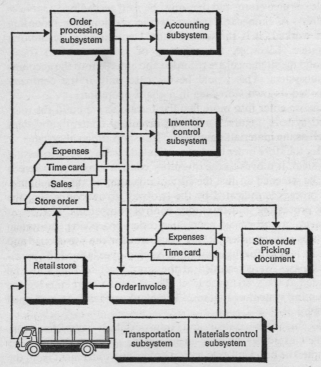

Figure 2 Addition of inventory control, materials control and transportation subsystems

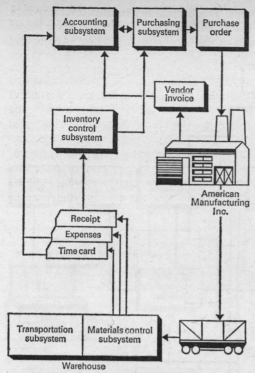

Figure 3 Adding the purchasing subsystem

produced by computer or can be done manually. The purchase order is made out and sent to the respective vendor or manufacturer. The vendor in turn fills the order, ships the product to the warehouse, and submits his invoice for entry into the accounting subsystem. This initiates Accounts Payable activity and the eventual reconciliation of cash payments. The two-way arrow indicates the interrelationship of the accounting and purchasing subsystems.

The blank card of Figure 2 can now be identified as a receipt card. This card enters the inventory control subsystem to update pertinent inventory records and is also passed through to the accounting subsystem to form the basis for vendor payment.

Figure 4 brings together the subsystems that have been mentioned up to this point. Two additional arrows have been added to the illustration. The first connects the inventory control and accounting subsystems. This is necessary because inventory data is a requirement of accounting reports. The second arrow shows the important output of the accounting subsystem. Accounting

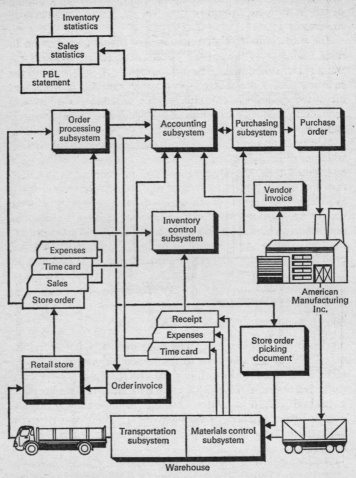

Figure 4 Basic subsystems

acts as the score-keeper of all the subsystems mentioned thus far. It accumulates data such as store orders, cash sales, expenses, receipts and vendor invoices, and from the basic source information, produces a host of meaningful management reports.

Three of these reports are noted in Figure 4 – profit and loss statement, inventory statistics, and sales statistics. These reports might show, for example, profitability by region, by store, or by department or might show sales by salesman or item grouping or might indicate inventory turnover or return on inventory investment.

The subsystems discussed thus far are basic ones common to most businesses. They are fairly routine computer applications, but it is still uncommon to find a computer system built with the required open-ended integrated approach that has incorporated all of the subsystems and still has the capability of additional more sophisticated subsystems. Figure 5 shows more advanced subsystems which can build on the basic ones already described.

Merchandising and management planning are the subsystems which produce the real payoffs from an integrated system. As a by-product of the basic subsystems, the merchandising subsystem uses the basic source data which has been collected and filed. However, the data is processed in different ways to answer specific management questions – questions such as the effect of promotion on sales, the effect of pricing changes and product mix, the comparative advantages of a limited as compared to a full product line, the significance of store layout and shelf allocation on

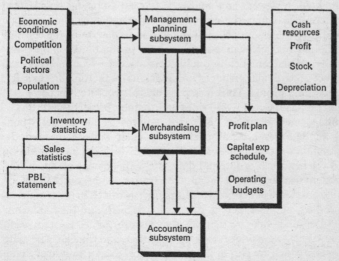

Figure 5 Addition of merchandising and management planning subsystems

profit, and the desirability of introducing new items or product lines.

The merchandising subsystem has been all but neglected by today's computer systems. There are several reasons for this. The first is the absence of the integrated concept. The subsystems previously described were not built to collect and store data for other uses – they were built only for the immediate job be it order processing or payroll. Therefore, the basic data needed for more sophisticated analysis is not available – it has been captured but not retained. A second reason why the merchandising subsystem is not often tackled is that the payoffs from this subsystem are not as tangible as from the other subsystems. It is easy to see the clerical savings from computerizing order processing but it is more difficult to see how a change in advertising policy can improve profit margins. It is not that the potential benefits of the latter are insignificant – to the contrary, the benefits can be far more meaningful than computerizing order processing. The point is that conservative management is less convinced that the benefits can be attained.

The management planning subsystem is the core of the integrated system. It is here where the major company decisions are made. Reports produced by the accounting subsystem are digested, reviewed, and analysed. Overall company policies are determined and promulgated. Sophisticated integrated systems employ business models and simulation to determine the effect of different management policies on the profitability of the company. A mathematical model is constructed which indicates the effect of particular management decisions on factors such as sales, costs, and profits. Thus it is possible to structure a paper world to react as the real world does and thereby project the results of management decisions before they are made. The model is built from data captured as part of the basic subsystems and the simulation is conducted using the same historical data. In addition to the historical or internal data, the management planning subsystem uses external data gathered from outside sources. On one side are the basic economic factors such as trends in gross national product, political factors, population growth and the like, while on the other side are the available cash resources such as profit plow-back, new stock issues, depreciation, and the like. These are factors which are part of the mathematical model built to evaluate and determine management policy. The results of the management planning subsystem are merchandising policies as well as the profit plan, capital expenditure schedule and operating budget. These figures enter the accounting subsystem to form the basis of measuring actual operation. They form the yardstick for measurement. The cycle is thus completed. Figure 6 presents the complete integrated system that we have been building in stages.

The central data base

The development of the integrated system illustrates the need for gathering basic information from key source documents and using this information throughout the system. Likewise, it is equally important to develop basic master files, which are used in common by the various subsystems. These common master files are normally referred to as a *central data base*.

The central data base holds all relevant information about a company's operation in one readily accessible file. The file is arranged so that duplication and redundancy are avoided. Information concerning on-going activities is captured once, validated, and entered into the proper location in the data base. Normally the central data base is subdivided into the major information subsets which are needed to run a business. These subsets are:

1. Customer and sales file.
2. Vendor file.
3. Personnel file.
4. Inventory file.
5. General ledger accounting file.

The various subsystems use information from the same file. The key element in a data base concept is that each subsystem utilizes the same data base in satisfying its information needs. Duplicate files or subsets of the central data base are eliminated.

The data base concept is a most logical approach to the paper work explosion which has hit many businesses. However, management should not overlook some important considerations in developing the data base as part of an integrated system.

A fairly obvious problem is the fact that erroneously entered data has an immediate influence on other subsystems and departments utilizing the data. Of course this can be a desirable factor in that the errors will be noticed more quickly and that feedback should help purify the data and ensure that necessary input controls are established and adhered to.

Another problem in building a single central file is the

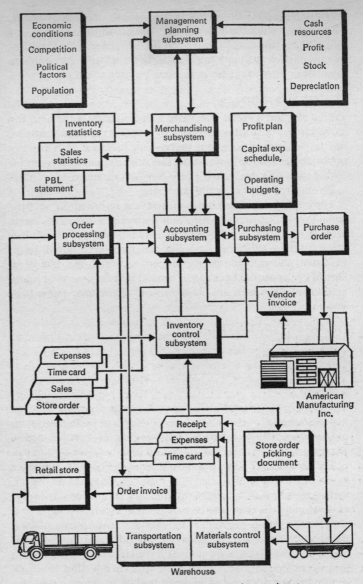

Figure 6 Integrated management information and control system

interdepartment cooperation that is needed to arrive at the pertinent data elements in the file. One department may need a degree of detail which may burden the reporting source such that the quality of all input suffers. The solution to these problems normally means discussions, meetings, and eventual compromise if the implementation time frame is to be met.

An interesting organizational phenomenon is often encountered during the development of the data base. This phenomenon can be called the 'geometric organizational syndrome'. It stems from the fact that joint decisions are needed to reach compromises related to the central file content. This does not seem a formidable task unless one realizes that the interaction pathways of four people are more than double that of two people. The progression is more of a geometric nature rather than an arithmetic nature. Thus, there is one communication path with two people three with three people, six with four people, ten with five people etc. Thus, adding a fifth communicating party increases the communication pathways not by 25 per cent, but by 67 per cent. The 'geometric organizational syndrome' is probably even more accentuated by the psychological and political blocks that individuals bring with them into the situation.

Another problem in developing a central data base is information security. When individual departments maintain their own data files, security of file information is not a problem. They know that other departments will not have access to confidential facts about their operation. In a highly centralized organization, the confidential nature of the central data base is not of primary concern, but in a highly decentralized organization where divisions are autonomous, the centralization of data in one central file can represent a serious obstacle. The divisions are skeptical about the information they submit – they wonder how it is going to be used. They do not want their performance figures to be known by other divisions.

The suspicion generated by the security aspects of the central data base can have a significant effect on the validity of what is supplied by the various users. It is true that certain quantitative data can be checked for obvious errors, but there is little or no check on the qualitative data – and this data can be extremely significant in the planning and analysis functions. Qualitative data can take the form of how the customer is currently using the product or how he plans to use it in the future. It also takes the form of information on customer complaints and problems in an effort to improve the product and service. The real measure of the success of the data base concept is the input going into the system. The psychological considerations of submitting confidential data must be carefully considered.

Problems might also arise from the fact that there is a time dimension to information as well as a content dimension. For example, in a particular situation, it might take two to three months from the time an order is placed by a customer to officially validate the terms of the order. From the viewpoint of the sales office, the sales effort for this particular order is completed. They would like to see the results reflected in the particular period in which the customer placed the order. Their thinking is, that in ninety per cent of the cases, the contractual negotiations are a formality and do not change the basic terms of the order. This may not be the feeling of the production or accounting departments. Each of these departments has a different date at which time they would like to see the order reflected as part of the customer and sales data base supplying reports to their departments.

Unless the new system satisfies the requirements of each user, the user is compelled to maintain his own system which of course, defeats the purpose of the central data base concept. The longer the user runs his old system in parallel, the more difficult it is to get him to supply accurate data to a new system which is not serving his needs.

Instituting an integrated system built around a central data base system brings to light deep-rooted management problems – problems that would continue to distill and grow if left unattended. These are the 'quiet problems' – the problems of company communication, organization, and control. The development of the data base acts as the catalyst in bringing the 'quiet problems' to the surface.

12 P. G. Thome and R. G. Willard

The Systems Approach:
A Unified Concept of Planning

Excerpt from P. G. Thome and R. G. Willard, 'Aerospace management',
General Electric's Aerospace Group, 1966, pp. 1–13.

A disposition to economize available resources has historically
been among the lesser attributes of man. Apparently, he has
survived for millennia behaving like a profligate parasite, consuming resources at an increasing rate in order to fulfill his
growing ambitions.

With the growth in the number, the physical dimensions and the
complexity of his ambitious undertakings, has grown also his
capacity to count, to measure, and to project his needs. And the
development of these latent capacities has recently led to an
apprehensive conclusion – that all resources at his disposal are
finite, and that they must be managed through a new discipline
founded on farsighted techniques, rather than by the traditional
measures of expediency that solved problems in a fragmentary
way.

Projects of growing complexity and size, such as bridges,
skyscrapers, ships and communication networks undoubtedly
served to widen the vistas of visionary engineers for problem
solving. But it took the pressures of the crash missile-development
program in the last decade to identify the systems approach as a
far-sighted and total philosophy for defining and solving problems
of a national scale and importance.

Most practitioners of the related disciplines associated with the
systems approach would be the first to warn that this approach
is not a magic cure-all; they would hardly agree completely on
its definition, nor would they agree on whether it should be identified as systems engineering, systems analysis, or systems approach.
In fact, in most cases these terms are used synonymously.

The authors of this article contend that 'systems approach' is
a more descriptive and appropriate term – by its broader meaning
– than systems analysis or systems engineering. The reason is
that 'engineering' has a hardware connotation, and 'analysis'
implies the separation into parts or elements, of something which

is already in existence. The chief attribute of the systems approach is that it can be applied with equal effectiveness to problems of a material nature – with well defined boundaries – as to the very process of delineating the boundaries of a dilemma so it can be focused sharply into a problem and then be solved.

Orderly way of appraisal

The systems approach is an orderly way of appraising a human need of a complex nature, in a 'let's stand back and look at this situation from all its angles' frame of mind, asking one's self:

1. How many distinguishable elements are there to this seeming problem?
2. What cause-and-effect relationships exist among these elements?
3. What functions need to be performed in each case?
4. What trade-offs may be required among resources, once they are defined?

We said above that the systems approach is suitable for assessing the challenges of a broad scale or of national significance. In the aerospace arena, challenges of this magnitude as a rule are programs characterized by the following orders of difficulty:

1. A complex goal (involving a major system composed of hardware, computer programs, facilities, personnel and data).
2. A constantly changing environment (which affects objectives, constraints and criteria).
3. Limited resources for advanced development (money, manpower, facilities and time).

For programs of this nature, the application of the systems approach early in the conceptual phase helps to reduce the chances for oversight, or the occurrence of so called 'appraisal gaps'. And this is achieved by using a structured technique to continuously identify and assess the impact of changing objectives, constraints, and design criteria on required resources: technologies, personnel and facilities.

General guideline

If it were possible to prescribe some basic precepts to be used in developing total system requirements, these would be:

1. Start at the highest and most general echelon of cognizance and authority to determine the boundaries of the overall system.

2. Proceed to define the system, in stages of increasing detail, translating functional requirements into hardware requirements.

3. Don't prejudge solutions! Any solutions in mind should serve as guides, rather than points of departure, in the planning process.

To understand what the logic of the systems approach is, it is equally important to understand what it is not. In a sense, the logic of the systems approach is the converse of the logic used in inductive reasoning. The inductive approach (Figure 1) begins with particulars, i.e. the collection of data used for synthesizing a theory or plan. In this approach, the objectives, or the general statement of the problem serves mostly as a constraint – to limit the introduction of data relative to the problem. The inductive approach is sometimes used by functional organizations to assemble an advanced technology development plan. In such cases, proposed development tasks and programs are submitted separately by each of the technology areas, then selected and integrated into an overall program. As a result, the inductive approach often leads to *solutions looking for problems to solve*, instead of focusing on the utilization of development resources.

Logic of systems approach

Briefly stated, the logic of the systems approach is 'deductive/inductive,' and here's why. The thinking path provided by this approach is in the form of a closed loop that has distinct stages for timely inputs and continuous feedback. As such, thinking evolves in cyclic fashion, as it progresses from general objectives to plans (deductive process), then back to refining objectives and to detailing plans further.

Up to this point in this article we have discussed the basic philosophy of the systems approach; to discuss the stages of analysis involved in this approach we need the assistance of successive diagrams to develop and expand our concept in progressive stages.

In its most basic form, the concept of the systems approach is presented in the lower half of Figure 1, showing that 'objectives' are used as a point of departure. Then the objectives are analysed, by successive stages, into detailed requirements and approaches. In each stage, additional new data is introduced (inductive process) to support the analysis. Next, selection criteria are determined, compatible with the original objectives, and are applied

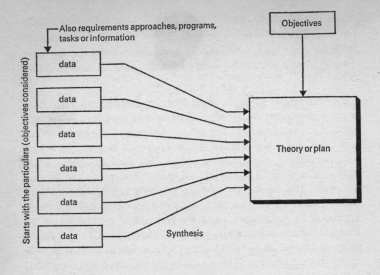

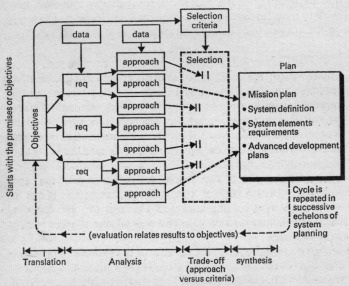

Figure 1 Planning approaches, the inductive approach, the systems approach

to choose the approaches; then the selections are synthesized (inductive process) into an advanced development plan or system design.

Importance of timing

The timing of contributions from specialists is a very important aspect of the systems approach. As considerations progress from the general objectives toward detailed requirements, the knowledge and judgment of experts in various technical disciplines are brought into play. But a significant point is that preliminary requirements must first evolve to a stage of maturity where needs can be defined in terms of functions. If detailed knowledge were tapped too early in the process, there would be a danger of formulating advanced development plans that are generated primarily by preconceived solutions.

Another cardinal rule to follow in implementing the systems approach is to practice disciplined documentation from start to finish. Thoughts are volatile, memories crowded, and organizations fluid. For these reasons it is vital to record the course of development, the approaches which have been considered, and the reasons for selections and rejections – a good plan must have the strength of a good argument.

We can gain further insight to the systems approach by viewing it as a *cycle* which has an expanding phase (the analysis) and a contracting phase (the synthesis). During the analysis the objectives are resolved into their constituent requirements and possible approaches; during the synthesis the approaches are weighed and selected through a trade-off study, then integrated into a system model or development program. This cycle *is repeated* at successive decision-making echelons of system planning or design, so that the output of each cycle is used as an input for the succeeding one.

To get closer to the structure of the systems approach, let's refer to Figure 2, which highlights the four principal stages or steps involved: translation, analysis, trade-off and synthesis. The translation, or initial formulation of the problem is an important step since it sets the course of all the work that will follow. Translation includes the interpretation of objectives and all recognized constraints on the problem solution. At this step in the cycle, the selection criteria are also determined. These criteria are later used in the trade-off study. Some categories of constraints, such as timing and policy, are also used as selection criteria. The

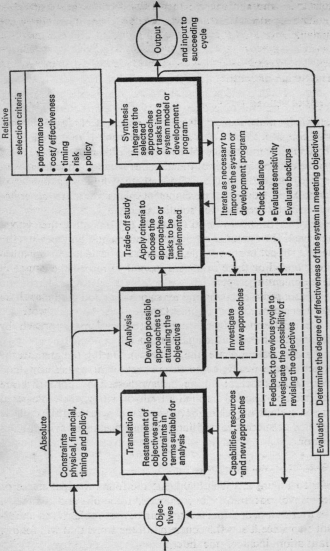

Figure 2 Steps within systems approach

difference between these two uses is that constraints are generally applied as absolute limitations, whereas selection criteria are applied later in the cycle to determine the relative merit of possible approaches. During the cycle a number of feedbacks may be required, as shown in the diagram, to improve and re-evaluate the output.

Decision-making cycles

One of the reasons why the systems approach is effective is that it is characterized by structural consistency. That is, the cycle described above recurs at each echelon of decision making – from the policy definition level down to the level of implementation, as schematized in Figure 3. We see here that at each echelon the possible approaches are evaluated against criteria, approaches are selected, and requirements determined. These decisions become the inputs to the following cycle. Thus, with each successive cycle the definition becomes more detailed. Incidentally, the reverse of the process – starting at the bottom of the diagram and proceeding upward – would represent an inductive approach to the planning process.

At this point then, not to lose sight of the very purpose of the systems approach for the thickening forest of chartwork, we should restate that we are following a structured approach for generating advanced development plans.

For all practical purposes, the advanced planning process begins with the mission definition cycle, after broad objectives have been stated based on policy definition. The advanced planning procedure can best be followed through a matrixed sequence depicted in Figure 4, where cycles are identified along the top, and the steps within each cycle are shown in the left column.

Definition crystallizes plans

When the mission Definition Cycle is followed to its completion, the output obtained is the mission plan 7 consisting of selected missions and their requirements. The inputs to the system definition cycle that follows, are the requirements for one of the missions chosen for advanced planning (8).

The system and its elements are defined in Cycles II and III, but only as far as necessary to determine the advanced development requirements for technologies, personnel or facilities. In Cycle II the system is defined by determining its elements;

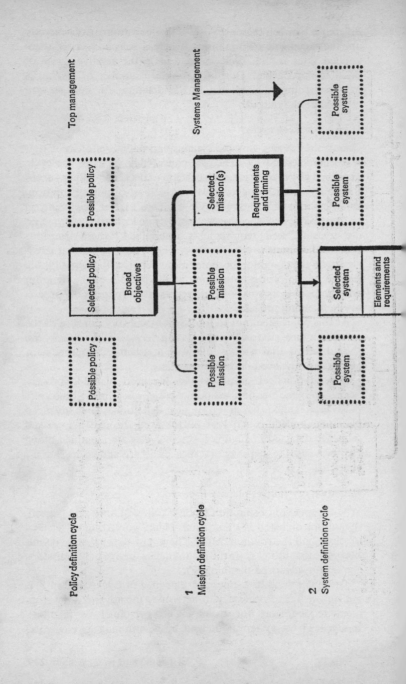

Policy definition cycle

1 Mission definition cycle

2 System definition cycle

Top management

Systems Management

Possible policy

Selected policy | Broad objectives

Possible policy

Possible mission

Selected mission(s) | Requirements and timing

Possible mission

Possible system

Possible system

Possible system

Selected system | Elements and requirements

Possible system

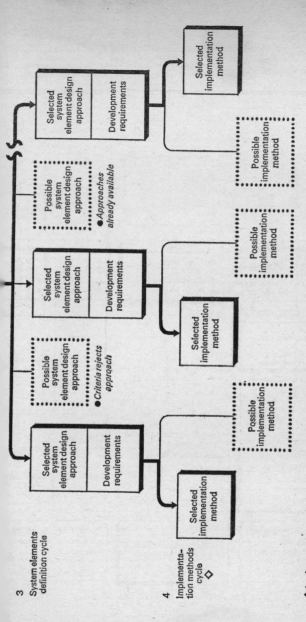

3 System elements definition cycle

Selected system element design approach | Development requirements

Possible system element design approach
• Approaches *already available*

Selected system element design approach | Development requirements

Possible system element design approach
• *Criteria rejects approach*

Selected system element design approach | Development requirements

4 Implementation methods cycle ◇

Selected implementation method

Possible implementation method

Selected implementation method

Possible implementation method

Selected implementation method

Possible implementation method

Selected implementation method

◇ Implementation methods include advanced development plans or end item requirements

Figure 3 The concept of sequence of cycles

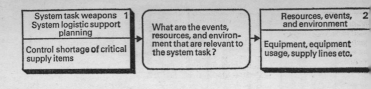

| System task weapons   1
System logistic support
planning

Control shortage **of** critical
supply items | What are the events,
resources, and environ-
ment that are relevant to
the system task? | Resources, events,   2
and environment

Equipment, equipment
usage, supply lines etc. |

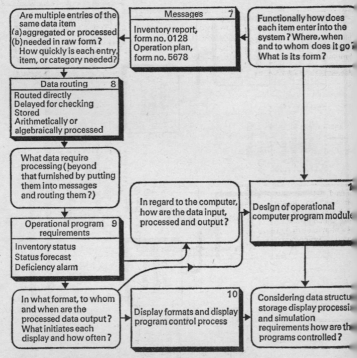

Are multiple entries of the
same data item
(a) aggregated or processed
(b) needed in raw form?
How quickly is each entry,
item, or category needed?

Messages   7
Inventory report, form no. 0128 Operation plan, form no. 5678

Functionally how does
each item enter into the
system? Where, when
and to whom does it go?
What is its form?

Data routing   8
Routed directly Delayed for checking Stored Arithmetically or algebraically processed

What data require
processing (beyond
that furnished by putting
them into messages
and routing them?)

In regard to the computer,
how are the data input,
processed and output?

Design of operational
computer program module

Operational program   9 requirements
Inventory status Status forecast Deficiency alarm

In what format, to whom
and when are the
processed data output?
What initiates each
display and how often?

10
Display formats and display program control process

Considering data structu
storage display processi
and simulation
requirements how are th
programs controlled?

Figure 4 Advanced planning procedure (for a product area)

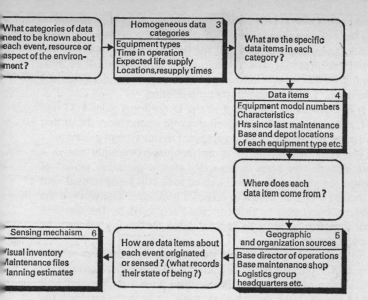

What categories of data need to be known about each event, resource or aspect of the environment?

→

Homogeneous data categories 3
Equipment types
Time in operation
Expected life supply
Locations, resupply times

→

What are the specific data items in each category?

↓

Data items 4
Equipment model numbers
Characteristics
Hrs since last maintenance
Base and depot locations
of each equipment type etc.

↓

Where does each data item come from?

↓

Geographic and organization sources 5
Base director of operations
Base maintenance shop
Logistics group
headquarters etc.

←

How are data items about each event originated or sensed? (what records their state of being?)

←

Sensing mechaism 6
Visual inventory
Maintenance files
Planning estimates

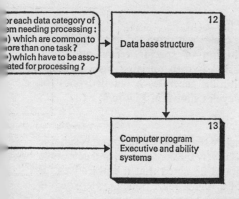

For each data category of item needing processing:
a) which are common to more than one task?
b) which have to be associated for processing?

→

Data base structure 12

↓

Computer program 13
Executive and ability systems

whereas in Cycle III the system elements are defined by determining their design approaches. The output of Cycle III is advanced development requirements (21) for the design approaches of the system elements for which development programs must be initiated. In other words, advanced development requirements are 'the needs' which exceed present capabilities.

It is essential to stress that in the Definition Cycles I, II and III, judgment must be exercised to avoid unnecessary detail that can obscure the overall vision essential to this type of planning.

The systems approach was not conjured overnight nor in a fortnight; it was born of necessity after a gestation period of almost ten years. It is a complicated tool, still being refined, and is used as a plan of attack to complex problems. Therefore, if the reader who is new to this concept has stayed on the tortuous trail thus far, he is to be congratulated for having reached the 'first plateau'.

So far, we have discussed the systems approach in general terms, as applicable to 'a product area'. To better understand the application of the systems approach as an aid to planning let's take a specific but compressed example, and assume retaining the interested reader's attention to the very end.

The example to be cited is a planetary exploration system that will be treated in the light of systems approach principles already described. Since the planning procedure for technologies, personnel and facilities (Cycle IV) are quite similar, only the technologies planning cycle will be discussed.

Particularization of procedure

A 'road-map' for planning cycles was provided in general terms through Figure 4; substitution of specific terms applicable to the planetary entry mission is then our first task. In fully-expanded form this substitution generates a detailed plan for each of the four definition cycles shown in Figure 4.

Mission definition cycle

To begin this cycle, we translate the objectives into terms suitable for analysis, identify quantitatively the constraints, and determine the selection criteria. It should be noted that the objectives for this cycle are the output of a higher level policy cycle which is the responsibility of the cognizant top management.

In identifying the constraints quantitatively it is not always possible to specify a numerical value with a high degree of confidence because of the many external contributing factors. Nevertheless, as John Stuart Mill noted in the last century, 'Knowledge insufficient for prediction may yet be valuable for guidance.' For instance, the total funds available for planetary programs over the next five to ten years may not be known with certainty. Hence, the proper approach is to specify the most likely available range of funds, in order to make it possible to understand the effect of such a variable on the conclusions, and thus to develop an approach (if possible) that is valid for the entire range.

Another consideration is whether to treat a constraint as 'physical' or 'financial'. In our example, launch vehicle availability is treated as a physical constraint, while it is obviously dependent to a certain extent on funds available. This is resolved by excluding launch vehicle development funds from funds available, and including only flight article procurement funds. However, it should be noted that placing of constraints in the financial area tends to broaden the scope of the analysis; this should be minimized if effective results are to be achieved.

The selection criteria are a means of measuring how well the possible approaches meet the objectives. There are two problems associated with characterization of selection criteria: the determination of appropriate criteria, and the establishment of their relative importance. For instance, criteria are derived from the objectives and constraints by asking such questions as: how well does the proposed set of missions match the scientific objectives? Is the emphasis on the proper place? Since the scientific objectives cover a wide range – extending from the exploration of Mars to exploration of comets and major planets – is the proposed set of missions broad enough and yet of sufficient depth to really advance our state of knowledge of the solar system? Are they consistent with NASA objectives, e.g. to maintain program continuity?

The establishment of the relative importance of the criteria can only be made after a detailed study of the objectives and constraints.

Here the analysis step involves development of possible approaches, within the constraints, in attaining the objectives. The procedure is to analyse the possible approaches in successive stages of increasing detail. For instance, the classes of possible planetary programs are first identified, then the classes of spacecraft which fit within each of these programs are developed. The classes of possible spacecraft must be identified to the extent that they can be compared with the overall objectives and determination made as to whether or not they are within the constraints. Therefore, the definition must include such items as weight, scientific capability, cost, and manpower requirements. It is then possible to combine these into any number of credible combinations. For the planetary exploration systems, for example, a total of twenty-four combinations could be identified.

Having identified the possible mission combinations, these must be evaluated in terms of how well they meet the selection criteria. Every mission combination must be compared with each of the others to assess relative values. This is repeated for each of the criteria and a total evaluation is obtained by combining the results for each.

It is always wise to assess the validity of the conclusions by conducting a sensitivity analysis. This helps to determine the effect of slight changes in each of the value assessments, on the overall conclusions. If the conclusion is sensitive to a small change in the value assessment, then further effort should be made to insure the exactness of the value assessments; or, the plans should be constructed so as to be as compatible as possible with the more likely mission combinations.

The output of the mission definition cycle is the identification of the missions for each year that best meet the objectives within the identified constraints. This is described in terms of target body, type of mission, spacecraft class, launch vehicle and the broad scientific objectives. In this example, the first opportunity for a sophisticated planetary entry and landing system is judged to be the Mars 1975 Voyager. Therefore, the next three planning cycles concentrate on the requirements for this Mars entry lander system.

System definition cycle

As in the mission definition cycle, the first step in the system definition cycle is to translate the objectives into terms suitable for analysis, identify quantitatively the constraints, and determine the selection criteria. More specifically, the objectives are translated into top-level functional requirements needed to implement the system for the mission. We also need to determine for each of the functional requirements the duration, the external environment during performance, and the reliability goals for each in order to meet overall reliability objectives. At this point, however, we come face-to-face with some constraints: external factors that influence the solution of our problem. For instance, the two chief physical constraints within which the planetary entry lander must be defined are: (a) the uncertainties in our knowledge of the planetary environment; and (b) the requirement that the entry-lander must be functionally compatible with the spacecraft that will deliver it to the vicinity of Mars. In addition, we must take into account the influence of financial constraints on the system definition. If the total cost of the lander exceeds the cost allocation from the previous cycle, the possibility of performing such a mission in 1975 is greatly jeopardized. Likewise, the selection criteria also are chosen to provide a measure of how well the possible systems meet the overall objectives. At the system definition level we see that the selection criteria range from reliability to growth capability for future missions.

The analysis step involves the application of constraints to determine the possible system definitions in successive stages of increasing detail. For instance, the procedure is first to analyse top-level functional requirements into more detailed requirements; second, to identify the system elements needed in implementation of these functional requirements; and third, to identify the system element performance requirements necessary to achieve the desired results.

The key to the successful execution of the analysis step is to judiciously choose the proper depth of analysis. If experience indicates that there is nothing new about a particular functional requirement, so far as implementation is concerned, then the analysis is not continued any further. We should keep in mind that the purpose of this procedure is not to design the system, but to identify advance developmental requirements.

Having identified the possible requirements for the functional and the system elements, we now evaluate them through trade-offs to determine how well they meet objectives. Again, in order to test the validity of conclusions, we need to perform a sensitivity analysis to determine the effect of small variations in the value assessment on overall conclusions. The output of this cycle is a list of all the system elements required to implement the system, the performance requirements for the system elements, and a mission profile which is a time history of all events and functions. Performance requirements are not necessarily identified for each of the system elements if it has been judged that nothing new is required to implement a particular element. In the same token, some of the requirements on the system elements, such as life, environment or cost may not have to be identified explicitly, if they are judged to have no bearing on advance development requirements.

System element definition cycle

In this cycle, we analyse the system elements that require further study. We need to determine their design requirements and design approaches at this time, so we can assess the necessity for advanced development.

To determine the system element design requirements, the elements are first structured into categories such as structures, retardation, data handling etc., for ease of analysis. Typical constraints identified at this level are overall lander weight, size limitations, as well as policy constraints prescribing the use of only flight-proven elements. Some of these trade-offs reflect that there are a number of approaches that could successfully meet the system element requirements. In this case, no further analysis is required, and that particular element is eliminated from further planning consideration. A situation frequently encountered in this cycle is that during a trade-off study we may realize that additional study is required before the possible approaches can be traded-off against the criteria. This could also occur in the other two cycles. The assembly of these study tasks that are required to identify proper design approaches is called a system study plan. We use this plan to map out the additional analyses and trade-off studies needed to identify the advanced development requirements.

The output of the system element definition cycle is a list of advance development requirements in terms of the design approaches for a specific system element, the performance requirement of these design approaches, and the timing as to when the technology must be available.

Technology advance planning cycle

Once the technology deficiencies have been identified for the system under study, plans must be developed to achieve this technological capability. The procedure for developing these plans is standard and straightforward.

Implications for the future

The systems approach, used essentially as described here, has proven an effective tool for planning advanced systems and technologies. When skilfully implemented, the systems approach provides a disciplined technique for:

1. Effective identification of advanced requirements for complex systems.
2. Thorough assessment of the effect of changes in environment on the development plans.
3. Timely identification of problems and study requirements in the conceptual phase of a program.
4. Accurate documentation tracing the chain of decisions and supporting reasons for communicating and justifying recommended courses of action.

Use of the systems approach in planning is justified in all types of applications where resources are limited and the systems are sufficiently complex that an intuitive or an inductive approach would lack the necessary thoroughness. The precision of this approach is limited only by the validity of the original objectives, the constraints, and the criteria, and the degree to which they can be quantified. In addition to being used in the aero-space business, the systems approach is presently being applied to such areas as oceanics, education, and water management. It is reasonable to predict that the applications of this approach to advanced planning will expand, and will provide substantiated bases for decision-making related to public programs and to the management of resources on a national order of magnitude.

13 J. Jaffe

The System Design Phase

Excerpts from J. Jaffe, 'The systems design phase', in Perry E. Rosove (ed.), *Developing Computer-Based Information Systems*, J. Wiley, 1967, pp. 94–137.

Organization of chapter

This chapter will describe system design by breaking it down into a number of more or less discrete processes, logically related to one another in that the set of design data from one process becomes a prerequisite for the start or successful completion of another. These relationships are shown in Figure 1 as a flow

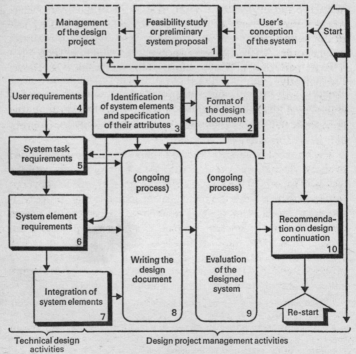

Figure 1 A representation of the system design process

diagram of the overall system design process. The major portion of this chapter will discuss, in turn, nearly all design processes, shown as numbered blocks in the diagram.

Each arrow represents the transmission of certain classes of data about the system being designed or *about the design process itself*. These data are transmitted from one process to the next. They must have specific origins, destinations and content. The design data that are created in one design process and then transmitted to another design process must be in such a form that they can be accepted without any ambiguity or uncertainty about their meaning.

Perhaps it might be well to state two of the things that this design representation is not. For one, it is not a complete representation of the design process; there is more to design than appears here. For another, it is not a complete time sequencing of the design processes. However, one may consider the representation to be roughly time-ordered, the processes starting earliest being at the top and continuing down to the bottom of the figure, where the processes represented are the end of some describable phase of the design effort. The numbering of the blocks also indicates a rough order. There are some processes that continue throughout a large part of the design effort; duration cannot be shown conveniently.[1]

The primary flow of data through the various design processes is indicated by the bold solid lines and arrows. The bold dotted lines show the primary feedback loops. Many other loops probably exist, formally and informally.

The horizontal brackets at the bottom of the figure are intended to show which design processes are concerned primarily with the technical aspects of the system being designed and which are concerned primarily with the management of the design process. It should be noted that there is a difference between the design process itself and the larger 'system' of which that process is but a part. In a sense one can say that there is a 'system-design system.' It consists not only of the management of the design processes and the technical work itself, but also of the entire environment within which the design effort proceeds.

1. Although design is accomplished by people, one should not infer from the diagram that these processes are performed by separate groups of persons simply because they are shown in separate rectangles; they may or may not be, depending upon the nature of the project management, the complexity of the system to be designed, the time available for design, and other factors.

The process blocks in the diagram are numbered for the sake of referencing; each topic will refer to one of the blocks. The ascending order of numbers is intended to represent the rough sequence in which these processes should be initiated in the design process. (The time dimension is also established vertically, as mentioned before, but the numbers of the rectangles help to resolve the position of processes represented on the same horizontal.) Unnumbered process blocks shown by broken lines are not discussed in this chapter.

An approach to design: the mapping of system tasks

The 'mapping' of a system that Beer talks about (see Echman, 1961) is accomplished by creating a sequence of means-end relationships which link the more abstract objectives of the system user to successively more concrete and specific goals. This takes place by discrete steps (the number of steps may not be the same for all parts of the system). During the development of this sequence of means-end relationships, the design statements begin to sound like descriptions of discrete 'tasks' performed by the system elements; an example for the taxi system might be plotting on a map the location of waiting passengers in order to determine which two can be picked up by the same cab without its deviating from a straight line.

A task is composed of a set of actions performed by people (physically or intellectually) and/or by machines in order to reach specified objectives. To control a fleet of taxicabs is an objective (a means for making money), but to plot the location of waiting passengers on a map is a task that is a step or series of discrete actions in the accomplishment of that objective. Tasks are in large part structured by the nature of the task environment. In the taxicab example, the manner in which the dispatcher plots the location of waiting passengers is in part determined by the quantity of geographical details, the precision required, the location of landmarks, and the scale of the map.

It is characteristic of objectives, on the other hand, that they are structured in large part by the values and judgment of men. A further distinction between objectives and tasks lies in the fact that objectives (of the user) vary, depending upon the nature of the environment, the level of the organization, and the characteristics of the individual user; objectives also are altered during the operation of an information system. The tasks, which are relatively concrete means, are performed in a less changeable fashion.

A major job of the designer, then, is to get the user to recognize the relationship between his objectives and the task mechanisms for accomplishing them. In order to do system mapping, not only must the means-end sequences be defined, but also the reasons for the existence of each step in the sequence and the explanations of how each particular means achieves each end. We will call this process *rational analysis*.

The first general step in rational analysis is the preliminary analysis of the user's requirements in terms of the objectives he wishes to accomplish. The second step is the decomposition of the objectives into subobjectives, at one or more levels, until the level of tasks has been reached. The third step is the successive refinement of tasks to the point where the designer can infer the data inputs, outputs, and processing the system will require.

When phrased in this way, it should become clear that the early steps in system design are the hard ones and that we do not yet have ready-made analytic tools for the job. There is little practical point in using precision tools when faced with system task requirements that are imprecisely defined. One would not think, for instance, of dissecting a whole elephant with a microtome. Therefore, in the section that follows, emphasis is placed on the process of making system task requirements more and more explicit. On the one hand, the requirements will be related to the ways in which they will attain the system objectives; on the other hand, their relationship to the hardware, software, and people that will perform the actions will be defined.

In the next section, comprising the bulk of this chapter, seven of the design processes shown in Figure 1 will be discussed in its turn, in accordance with the block number in the figure. The intent will be to identify and define the various design processes that contribute to the progress of system mapping and to the further explication of requirements and system actions.

The design process

Block 1 Feasibility study or preliminary system proposal

This is the first numbered block in Figure 1 and is the first explicit part of the entire design process. There are other ways of starting the system design process, some better and some poorer. A document that requests the initiation of a study makes a very good start. However, such a document should state the area under consideration and limit it in time, space, function, and structure. Alternatively, it should indicate in what areas no limits are contemplated. Frequently what is called for by this document is a feasibility study from which the system user can extract enough factual information to enable him to decide whether or not to continue with further design.

There is a myth connected with feasibility studies. This myth has very attractive features, for example, that one can specify through a feasibility study

1. A set of alternative design concepts.
2. The approximate length of time required to develop each concept.
3. The cost of executing each concept.
4. The amount of system growth each concept would allow.

Although believers in this myth admit that an almost complete set of alternative designs would have to be produced in order to yield the four specifications above, they reckon without the harsh world of reality that rarely, if ever, allows them more than a fraction of the amount of time and/or resources required for fulfillment.

What then should a feasibility study be? It should be a set of general statements that in effect present the opinion of the investigator about whether anything useful can be accomplished by further work. A good study document can be expected to point to the system under consideration and indicate the incremental utility of the next few steps toward a more definitive statement. It can indicate the general types of services the user expects to obtain from his system: for instance, whether he wants scientific computation, inventory control, large data storage, rapid access to information, sophisticated displays, or various combinations of these and other possible services. Such preliminary statements will do much to indicate the future course and cost of a system.

The situation in which there is no such feasibility document to start a major design project is rare. The reader might ponder for a moment the results of starting a design project without presenting statements of purpose, limitations, and interfaces relating subsystems to the system as a whole.

The results of a feasibility study provide the first statement of technical and managerial guidance to be used in making plans for subsequent design work. Where a previous feasibility study does not exist (produced and hoarded perhaps by a different agency), the design process is started by a preliminary system proposal. Either the proposal presents approximately the same content as the feasibility study and assumes that the accomplishment of a design is feasible, or it must lead to the undertaking of such a feasibility study. Without one or the other further design work cannot proceed rationally.

Block 2 Format of the design document

Usually the results of the design effort are presented to the user in the form of a document containing prose, block diagrams, tables, mathematical equations, and other modes of presentation. The format of this document must be specified in advance for the following reasons:

First, as we have already indicated, the substantive content of the design material is likely to be new when it does not deal with equipment specifications. We do not know the names of the design parameters (a characteristic element); this causes errors of omission. There is a tendency to use ambiguous terminology; this causes errors of commission. It is very easy to make such errors when there is no conceptual design schema to which to refer.

Second, when so much of what is designed is conceptual in nature, communication between the designer and the client poses a problem. The designer should show his customer the derivation of the multiple sets of relationships that will comprise his system, in a form that leaves as little room for individual interpretation as possible. From such a form, the reasonably knowledgeable reader should be able to trace the logic which led the designer to specify certain requirements, whether for data, organizations, system procedures, or hardware.

The format of the design itself (which is going to be a document) requires an analysis before it is determined. The first pieces of information needed to select an appropriate format are the uses to which the design is to be put. For instance, if the design is only preliminary, as a feasibility study would be, a rather simple narrative form may be adequate. If the design is to be used as a basis for specifying equipment characteristics, it has to be adequately partitioned so that statements relating to a particular piece of equipment can be associated with one another. Furthermore, the statements that contain the specifications must be couched in terms familiar to the equipment designer. If the design is going to an architect, it might be presented in the form of a floor plan and elevation or a set of numerical requirements, for example, to house a certain number of people, with a certain square footage per person, a specified number of operating areas, and so on. If the design will be used by a personnel manager, it must state the skill levels and specialties of the people needed to man the system.

The two things that all formats should have in common is an agreement on vocabulary between the designer and the receiver of the design, and the rationale whereby the design was worked out. In all cases the designer, before he starts his work, must obtain from his user or users a clear statement of the intended application of the design and an agreement in regard to the format, a common working vocabulary of terms, concepts, measures, or simply standards and conventions.

The design format should be specified precisely and agreed upon by designer and client. A clear format will significantly reduce the communication problem at all stages, from design concurrence implementation to system evaluation. To ignore this step is to risk subsequent misunderstandings among users, managers, and designers.

Block 3 Identification of system elements and specification of their attributes

What are the system elements to be specified? In designing a new radio, for instance, reference can be made to a number of books that specify standard elements, such as speaker, tuner, and amplifier. Each element is further described by a number of *attributes* (inherent characteristics). The speaker must reproduce a certain range of 'frequencies' and output each frequency at a certain sound level. The specification of elements and their attributes lies at the heart of any design process and is perhaps the logical starting point for it.

Attributes are the essential dimensions used to describe the system elements so that, if the attributes were missing, the elements would not accomplish their functions. When the system elements and their attributes are described early in the design phase, the designer and the user have a concrete base from which they can determine whether the services desired from the system are going to be provided.

The elements or subsystems of an information-processing system that are listed below are convenient for analysis; other people may name elements differently, but a particular system concept will not change substantially as a result of nomenclature. The system elements are as follows:

1. Data.
2. Personnel.
3. Display.
4. Equipment (that is, input, output, transmission, logical and arithmetic processing, and storage).
5. Learning.
6. Retrieval.
7. Programming.

Each element listed possesses a set of attributes that also have to be described still more concretely. Listed below and discussed briefly are the attributes of four of the seven system elements just mentioned. Although the list may not be complete, it is, hopefully, adequate as a 'starter set'.

Data element. In a very real sense a system user's need for information can be satisfied only by acquiring and transforming data, providing the information is the essential function of an information system. After the acquisition and transformation of data by the system, the data are further transformed into information by the user when he exercises judgment and formulates actions that will help to accomplish his objectives.

Actually, an information system may encompass several parts, people, or machines, each of which acquires and transforms data. Some of these parts function in series and some in parallel. Whatever way they function, and no matter how simple or complex the system may be, data ultimately are always transformed into information about one or more of the following states:

1. The environment with which the system and its users are concerned.
2. The resources that the user controls.
3. The information-processing system itself.
4. Commands to the system.
5. Commands to the user's resources.[2]

2. For analytic purposes the distinction between data and instructions must be constantly borne in mind. The data are what the user eventually transforms into information. Instructions are messages to the system or to user resources that initiate or terminate actions in specified ways. This is the same distinction that is applied to computer data, which must always be either data or instructions.

The attributes of the data elements are as follows:

1. *Relevance*. There must be a rationale which makes explicit the relationship between each class of data and the system task that requires it.

2. *Nomenclature*. The names of the classes of data, the units used, and the modifiers or qualifiers must all be expressed in a language common to the users and designers of the system. There must be no ambiguity in terms.

3. *Grain*. Attention must be paid also to the degree of fineness of the data; that is, the denotation of the most specific case in each class and each higher level of aggregation. For example, if the class of data is weapons, one must specify whether this has to be subdivided into manned aircraft and missiles, and these in turn broken down into their constituent types, B-52G and so on. In complex systems, different system parts may require different levels of aggregation.

4. *Sources*. For each class of data the place from which it originates must be specified. Some data, as previously mentioned, result from the cognitive process of the system users; in such cases a person or organization must be designated as the source.

5. *Destinations*. All data must be characterized in a system by destinations: the potential users may be either men or machines. Most often data will have a number of destinations in the form of intermediate processing points or stations.

6. *Logical relationship*. Some data in the system may be causally related to other data, some may be covariant, and some may be independent. This state of affairs frequently is seen in data generated by physical events, the status of an aircraft, for instance. The datum that an aircraft is capable of taking off is derived from a whole pyramid of other data reflecting such things as the state of the weather, the condition of the runway, the flight characteristics and physical readiness of the aircraft itself, and a ready pilot. These relationships must be specified so that when the data are processed the proper sets can be established. This is extremely important in computer programming, where failure to associate proper items can require extensive reprogramming.

7. *Processing*. Data entering an information system may

(a) be examined logically (e.g. sorted or examined for validity);

(b) be entered into a mathematical calculation (e.g. summed or used as a value in an equation);

(c) pass through the system essentially unchanged.

It must be determined whether each class of data will be either saved or destroyed, or whether a capability will have to be built into the system to give a person these two options. Each class of data in the system will require one or more of the three types of processing. Perhaps the largest part of the system design job is specifying the data-processing requirements.

The last statement would be true if we took into consideration only the processing of data required by the system user in pursuit of his objectives. When we add the processing of instructions (for example, retrieving data from intermediate storage, without which they could not be processed), the job becomes large indeed.

Processing, then, basically involves the mathematical and/or logical statement of how each class of data will be transformed and used by the next part of the system. In an automated information system the final processing of some sets of data takes place within a piece of equipment and is never apprehended directly by a person. The final processing for other data produces displays, a part of an information system which will be discussed shortly.

There are other attributes of data, such as their form and timeliness, but these are really more closely related to the attributes of displays. It is necessary to remember the existence of the interaction between data form and timeliness, on the one hand, and displays and communications, on the other; just such interactions require attention in the system elements integration effort. Data form and timeliness partially determine display equipment characteristics and communications load. The reverse is also true.

Personnel element. Various concepts of a personnel subsystem are prevalent today. Generally they include the attributes of the people who will operate a particular system.

In an information system people not only use the outputs of the system but also are components of the system. It is helpful to distinguish again between these two cases. In one situation people monitor and use information produced by the system but do not serve as an information input source (except very indirectly through policy and procedures) for the rest of the internal system. In the other situation people not only monitor and use the information but *do* inject instructions into the system in the form of computer requests; these are instructions to which the system must respond. The latter case still requires the precise specification of the attributes of the people and their actions that is customary in the former. There is a shift of emphasis, however, from equipment operation to data organization and retrieval. This is characteristic of information systems that require more complex man-machine interactions. These interactions are more flexibly structured than in the situation where men are more passive receivers of data from noncomputing machines that have a relatively limited repertoire of response. In both cases, however, the following are attributes of the personnel subsystem.

1. *Distribution of responsibility.* Primarily included in this attribute is the determination of the various 'seats' of decision-making prerogatives. This means the specification of the location, skill, and number of people or groups of people who have the responsibility and authority for taking all of the various types of actions available to the system users. *The authority, the responsibility, and the knowledge should reside in the same place.* Furthermore, there must be recognition of the fact that *too high a degree of specification and formalization will decrease the ability of the system users to structure and restructure their organizations and actions* to meet changing conditions. This attribute is closely related to the specific characteristics of the physical communication network, the latter being in part derived from the former. Organizational communication channels include not only who talks to whom but also how the hierarchical and lateral power relationships are structured. It includes who gives orders to whom, who can make requests of whom, and who passes information to whom.

2. *Operating procedures.* The human actions (as well as equipment and computer program actions) and the methods for accomplishing them are an inherent part of any information system. The human actions are as follows: monitor, compare, assess, predict, decide, command, inform, request and comply. These are the things people do in information systems. The particular applications of these actions, plus the ways in which they are accomplished, form the operating procedures. They range from the specification of the format of any communication to the operation of equipment. They also include rules, regulations, situation descriptions and criteria that guide people in taking actions.

3. *Extra-organizational actions.* The personnel in a system have two other types of activities in which to engage besides those directly connected with the operation of the system. They have to deal with people and agencies external to the particular system with which they are involved, sometimes on tasks not closely related to system tasks. They also need to be concerned with the maintenance of the human organization; this is particularly true of supervisors. By knowing of these extra-organizational activities, the system designer should be able to keep from overloading the human operator. Too frequently the designer assumes that the only thing people have to do is to operate the system.

Display element. Although each part of an information system has an output, the term 'display', as used here, refers to any presentation of data to people by means of equipment. Displays form the major interface between the user and his system. Because displays are considered to be both data and equipment, they must possess the same attributes that both of these elements possess. In addition, displays possess the following attributes.

1. *Degree of aggregation.* An aggregation is defined as a collection of like things. In an information system, the meaning is extended to include collections of data items at the same level of specificity. For instance, in presenting the results of a population census, a tabular display could show just total population (one cell), population by state (fifty cells), or population by congressional district (over 437 cells). This attribute is the one that partly reflects the distinction between command systems and control systems. As we said earlier in the reading, the level of summarization and specificity of the displayed information is dependent in part upon the nature of the system. A command system for higher echelons generally requires more summary and less specific data than a control system; an executive or commander finds it difficult to deal with a large number of discrete events and items. The data have to be aggregated or summarized in some way even though the final aggregation or summary may in itself be considered a single item in another whole set of related items.

2. *Coding.* This attribute of displays is well known to the human engineer. He is concerned with the particular form into which the data are placed to present a signal to a human being. When the visual modality is used, data are frequently coded in graphic or

pictorial form (maps). The other end of the dimension is coding which essentially is straight English narrative. Tabular formats probably are somewhere between the two extremes.

A great deal has been written about coding. In information systems, however, no great body of knowledge about it has been built up. Systems users and designers alike feel that for the performance of certain tasks, some data formats are more suitable than others. On first examination it would seem obvious that more information can be obtained from data coded in one form than in another, but it is very hard to demonstrate this except by using a specific example. For instance, in presenting data about the geographical relationships between two cities, we immediately think of using a political map. If we are interested in the terrain between the two cities, however, we might prefer a topographic map. On the other hand, if the datum needed is the distance between the two cities, a simple numerical format would be more appropriate and potentially more accurate. The presentation of distances between a set of cities could be accomplished by a simple numerical table. Suppose, however, that some data are needed in regard to the relationship among three or more cities. We might wish to know whether, in following a great circle route from city A to city B, we would pass through city C. This bit of information is very easily obtained from a globe or from an appropriate two-dimensional projection. It can also be obtained through mathematical formulations but probably less efficiently unless the data are going to be fed back into a machine computation.

Another example that occurs frequently in information systems is furnished by the need for predicting the status of aggregated events such as inventory items. In some cases it is traditional to present the information in the form of a frequency polygon or smoothed line graph. The latter makes it relatively easy for the user to infer from an extrapolation of the curve what the status will be at some future time. This item of information can likewise be presented quite readily in numerical form derived from an appropriate mathematical equation. If the information needed is the rate of growth or decay, a smoothed line graph gives an immediate indication, particularly in the case of a complex growth such as negative or positive acceleration. If, however, the function is a straight line, one might obtain a more precise indication of the rate by presenting the slope of the line in numerical form.

The examples just presented point out the need for very close examination of the information and decision-making requirements of the system user. This is important not only from the point of view of the people coding the data for most effective decision making, but also from that of the designers of computer programs and display equipment, the characteristics of which will be partly determined by the display requirements.

Let us go back for a moment to the statement, made earlier in the chapter, that command systems usually require more summarized data than do control systems. This statement should not be interpreted to mean that there is a need for highly coded data. In fact, while executives and commanders usually deal with highly aggregated data, there need be neither a high degree of coding nor preliminary interpretation or filtering by other agencies, human or otherwise. As a matter of fact, observational evidence indicates that in many problem situations decision makers should have access to essentially unfiltered, although perhaps aggregated, data.[3] Insofar as possible these data enable decision makers to draw their own mental picture of the state of their resources and environment. The so-called 'big picture' is too frequently interpreted by the system designer to mean that the executive should be supplied with highly coded and/or highly summarized data. From these the decision maker well may find it difficult to infer the nature of the objective state because it has been filtered in such a fashion as to reflect the thoughts of other people through whom the data have passed, these thoughts now forming an indistinguishable part of the data.

3. *Timeliness*. This is a major attribute of many elements of a system. Timeliness may be set by the nature of the events in the physical world which occur or reoccur at certain times and have certain durations. For instance, if event A takes ten minutes to occur and cycles continuously, data about event A more than ten minutes old are not timely, although they may be useful as history.

Another major consideration, along with timeliness, concerns the degree or kind of significant change. Event A may occur or reoccur every ten minutes but may not result in a change of status, or it may result in a change that is insignificant in terms of some criterion. Some thought will reveal the import of this subject in determining the frequency with which facts about the occurrence

3. Aggregating data entails a certain amount of error which can be considered filtering. Rounding errors are one example.

of events should be reported into the system and the criteria whereby cues for their reporting are established. 'Reporting by exception,' for instance, reduces the number of reports; reports are generated only when a criterion range of values is exceeded, or at infrequent periods to ensure the recipient that the reporting apparatus is still working.

The other aspect of timeliness has to do with the needs of the user rather than the nature of real world events. Information needs must be analysed in terms of the times at which the decision maker wants the data, the time he can afford to wait for them if they are not already present, and the cues whereby the needs for timely data are recognized.

4. *Initiation or access.* After we have specified the system task requirements, the data, the methods for inputting data to the system, and the processing of data, we generally tend to forget that the data will not just automatically be displayed. A mechanism must be provided for triggering the action of a part of a system to produce displays. The mechanism may be an executive computer program that governs the preparation of data for display and contains a set of instructions which enable it to recognize the appropriate moment to begin the production. This may be an internal clock, if the display initiation is completely predictable and cyclic, or it may be the recognition of a queuing event.

Another initiation mechanism may be a human operator, who then must be provided with a communication channel for cueing the display-producing part of the system. In a computer-based information system the generic term for this mechanism is a *display request device*, which may be structurally identical to various system input devices.

5. *Transduction.* We have borrowed an engineering term here to indicate that transduction is an interface between user needs and hardware requirements. The specification of this attribute is done in the naming of the type of mechanism (lantern slide, cathode ray tube, computer line printer etc.) for converting data contained in a form not recognizable by human beings. This specification is dependent not only upon the nature of the rest of the structural system but also upon the types of formats needed. For instance, it is very difficult to produce line graphs on a printer. Usually, however, there is a wide choice of alternative transduction mechanisms. The selection is frequently made on the basis of economic and structural engineering factors, such as the degree of optical

resolution required, the number of people who have to view the display simultaneously, or the updated speed.

6. *Correctness.* This is another highly important attribute from the point of view of the decision maker, whose performance is degraded by the presence of incomplete or incorrect data that are not known to be faulty. Simulation research has shown that major decrements in decision-making effectiveness are due to uncertainty. Uncertainty can mean not being sure of the true nature of the objective situation. It can also mean not knowing the completeness and correctness of the data with which one has to deal. The system designer should be sure to provide mechanisms which indicate whether there are significant variations in either of these two attributes of displayed data.

Equipment elements. The equipment element is composed of engineered units, the physical devices through which all data flow. When we are dealing with hardware design, it is convenient to regard the element as two major units, the computer unit and the communication unit. The digital computer is perhaps the more complex and, in a very real sense, is a complete system itself, one which inputs, stores, processes, outputs, and transmits data. However, we will not discuss the attributes of a computer as an item of equipment since we expediently claim that the information system customarily would employ a general-purpose one which can do almost any kind of logical and arithmetic processing typical of the class.[4]

For the analysis in this chapter we will consider only the communication unit. This unit includes all the hardware for sensing, inputting, outputting, and actually transmitting data through the system as a whole. The attributes of the equipment elements are as follows:

1. *Adequate stimulus.* An adequate stimulus will cause an event to be sensed or an impulse to be initiated and propagated. The hardware has to be capable of detecting the events of interest to the system.

2. *Stimulus production.* Although this attribute may be part of the previous one, special consideration should be given to whether the

4. One might want to remember, though, that computers do vary significantly in regard to such attributes as speed of basic operations, size of central memory, and parallel input, output, and/or computation processing.

adequate stimulus is provided by a human being or by an automatic device such as a radar.

3. *Data form and transduction agent.* In this case we are interested in the carrier of the data. It is a series of electrical pulses varying in frequency, amplitude, or other physical form. We are also interested in the mechanisms whereby its form is converted and reconverted throughout the physical system from origination to destination. Displays are a subset of specific transducers. So also are input devices by which the system users introduce data into the system.

4. *Capacities.* Here must be specified the rate or rates, under varying environmental conditions, with which data can be transmitted; capacities are expressed in units of data per unit time. Not only are they expressed for various environmental conditions but also for peak and steady situations and for both individual equipment units and aggregates of like units. Capacities are also characterized in terms of *duration* of either peak or steady performances without degradation, and in terms of reaction time from stimulus onset to system response.

5. *Compatibility.* This attribute must be considered in an engineering and human engineering sense and also in terms particularly applicable to information systems.[5]

6. *Reliability.* This is a standard attribute of equipment specified in most engineering plans. Reliability is a measure of the extent to which a piece of equipment or a system will respond to the same set of inputs in the same manner over a number of occurrences of these inputs. Reliability of data is a particular problem for information systems because of the number and complexity of transformations to which they are subjected.

7. *Adaptability.* This attribute is poorly defined at present but deals with such things as increasing capacity by additions of more of the same components or the insertion of new components without seriously interrupting the system operations and without performing a major redesign of any substantial part of the system.

5. In an information system compatibility can be defined as being able to accept the inputs from, and give outputs to, other components at a rate adequate to prevent queues or loss of data. When the output rate of one unit is much faster than the rate at which the next unit can accept the data, buffering is required to keep the data from being lost.

8. *Discrimination.* What is the minimum signal/noise ratio both from a signal transmission point of view and from a human engineering point of view?

9. *Faithfulness.* This is the specification of the ability and characteristics of equipment components, particularly sensing and receiving elements, which enable them to minimize (1) the acceptance of something as a signal when indeed it is not a signal, (2) the rejection of something as not being a signal when in fact it is a signal.

10. *Geographical distribution and number.* These attributes are shown by an actual map or schematic drawing of the arrangement of the major equipment units.

Block 4 User requirements

The design of information systems starts with the specification of the user's requirements and continues with an explanation of how the proposed system will meet them. Since this is initially a question of management's interpretation, this process is shown to stem directly from the management process.[6] Management must study the user's initial perception of the system and attempt to *forecast* the objectives that the system can help to accomplish when it becomes operational. The reason that this process stems directly from the management process block is that only management has access to the relevant data; once these data have furnished guides and limits to the technical design effort, only management can obtain the user's concurrence in the statements of boundaries and limits. Otherwise, as many of us have painfully discovered, the various costs of the end product will exceed the resources of the user.

Statements about the general objectives of the user should be phrased so that the designer can infer the kinds of tasks the system must perform in order to accomplish these objectives. Sometimes, of course, the tasks to be performed are defined as the objectives. This should not be permitted. Objectives are goals, and tasks are ways of reaching goals. Another way of expressing the same idea is to say that objectives are specified by asking 'why' and tasks

6. It is difficult to define here the level of management that should be involved; the level will vary partly as a function of the magnitude of the project. Management will have to be involved, however, because such personnel will present and interpret the design to the user.

by asking 'how'. Information systems can have only three general objectives:[7]

1. To communicate.
2. To process data.
3. To control environment and resources.

It may help if we consider that the last objective implies both of the preceding ones. Thus, a control objective implies the need for data processing and communication. Perhaps this sequential relationship is derived on the basis that when the last objective is chosen the others become means to the end.

Frequently some user objectives are so specific that they become a direct part of the system task requirements. Actually, then, these are not objectives in our use of the term; but it is well to remember that, if the user states that one 'objective' is to use only a Burroughs adding machine for any data processing, this is a design constraint and had better show up in the specifications. If the designer forgets to include it, he looks foolish for having constrained the design without apparent reason. Although such constraints help to limit the design, they must not be mistaken for objectives.

The user's resources in terms of time, money, and adaptability will also constrain the design. Adaptability is considered in the sense of the degree of radical departures from current systems that the user will allow in the development of novel concepts and the procurement of new items. However, constraints can be very helpful to the designer since they provide limits against which to evaluate the acceptability of the design.

One can also consider as a part of the user's resources willingness and ability to furnish data about his plans and operations as well as operationally experienced personnel to aid in the design effort. These become important factors because of the designer's needs to understand the user's operational objectives both in and of themselves and within the larger context of the user's general procedures and philosophy.

If the design process is to be iterative, the statement of the user's requirements, that is, objectives and resources, becomes a primary input to the decision on design continuation, which is described in block 10. It may well be that the preliminary iteration of the design indicates the need for a more extensive system than was

7. Pursuing the 'why' of a system ultimately leads to teleology. Since the focus in this book is on design of information systems and not on philosophy, let us stop at the present level.

originally considered. If the user's resources cannot support such enlargement, then the next design iteration will have to reduce system scope or sophistication. Or it may be that the increment in utility that the system would contribute to the user is proportionally small in comparison to the resources allocated. In any event, these problems had best be brought to light before implementation is begun.

Block 5 System task requirements

In determining the system task requirements the system designer must ask and answer the following questions:

1. What are the tasks?

2. Why is each task performed? (Means–end; higher-order means –higher-order end.)

3. Where is each task performed, and where are the resources which are manipulated?

4. By whom or what is each task performed?

5. With what is each task performed? (Data, computational routines, formulas etc.)

6. When is each task performed, that is, under what conditions, at what times? In short, what happens to initiate, continue, terminate, or reiterate the task performance?

7. How is each task performed? (This is a fitting together of the previous material.)

Systems are designed rationally when these seven questions are answered. The first four questions must be emphasized at the earlier stages of determining the task requirements. Each system task must be related to the supra-ordinate task. In short, *the relationship between lower-order tasks and the higher-order ones from which they are derived must be made explicit*. Many times this is not at all easy to do because of logical complexities, vagueness of statements, or implications potentially undesirable from either the designer's or the user's point of view.

There is also a need for trying to establish conceptually broad statements of system tasks so as to make the job of subsequent analysis easier and comparable across systems or across related work efforts within the same system. A set of such tasks is enumerated below; hopefully, they represent general categories. An information system will usually be required to accomplish a number of these tasks. A system may contain any combination thereof. It is one of the earlier and major jobs of the designer to select the appropriate set from this list for the particular system under consideration.

1. Planning to obtain resources.
2. Planning use of resources.
3. Assessing the environment.
4. Assessing one's own resources.
5. Manipulating or moving one's own resources.
6. Assessing system status.
7. Changing system status.
8. Interfacing with other systems.
9. Surviving (protecting the system's capability to accomplish its mission).

The series of tasks just enumerated is the one by which the user's objectives are accomplished. However, there will be differences in the performance of these tasks, depending upon whether the user desires to emphasize command or control objectives with his system. Command objectives are those in which the user is concerned mainly with planning for and maintaining his administrative and organizational mechanisms in good working order in a slowly changing environment. He is equally concerned with the relatively long-range capability of his resources, whether they are combat aircraft or supermarkets. Control objectives have more the connotation of directing relatively discrete, short-term movements of resources in a rapidly changing environment in order to implement command objectives.

Although the nine general tasks just enumerated apply equally to the satisfaction of both control and command objectives, the distinction can help the designer to determine what kinds of data the user needs to meet each type. A comparison of the data needs for command and control objectives is given below.

Command objectives	*Control objectives*
1. Aggregated and summarized data.	1. Discrete and specific data.
2. Speculative data (e.g. hypothesis and trial solutions).	2. Factual data.
3. Long-range predictions.	3. Short-term extrapolations.
4. Direct communication with other people.	4. Direct access to a data-processing element.

In summary then, the general distinction between the two types is that control objectives usually require a shorter system response time than do command objectives. The trouble with this distinction is not only that the boundary becomes hazy, but also that some so-called command objectives require short response times. Conversely, many so-called control objectives require look-ahead features which seem to satisfy command objectives. The distinction can be useful for initial matching of system capabilities with requirements, but persons interested in information systems should be very careful not to be snared by distinctions which mask the common nature of such objectives and which, if persisted in, can lead them to ignore a solution to a novel system problem simply because the solution mode is not characteristic of the particular system objective.

Nevertheless, a great many problems in system analysis and design have been caused by the failure to distinguish between these two types of system objectives. For a system user concerned with the task of planning to obtain resources, few data are more useless to him than the current status of each item of his total resources. He needs basic data about what the environment may be at a time which is far enough away for him to put into effect a plan for implementing changes in his resources.

The user objectives, then, help to provide a framework within which both the general and the specific nature of the system task requirements can be enumerated. Once the designer selects the appropriate set of tasks, he can derive from them successively more concrete subtasks. He continues to analyse these subtasks until he arrives at some level of description from which he can infer the nature of the mechanisms needed for the implementation of the tasks.

A set of diagrams such as those illustrated in Figures 2 and 3 should emerge from this analysis of system tasks. These figures present a simplified example of the analysis of a deterrent weapons information system. Figure 2 shows the most general level of analysis. It identifies the major tasks and their inputs and outputs. Each connecting line represents the flow of data to or from a task, culminating in the issuance of orders to the deterrent weapons force in the field.

Each numbered block represents a data-processing activity (in the general sense of the term) which may be quite complex and involve both computers and people. Roughly the same set of activities occurs in each block as follows: formulation of the

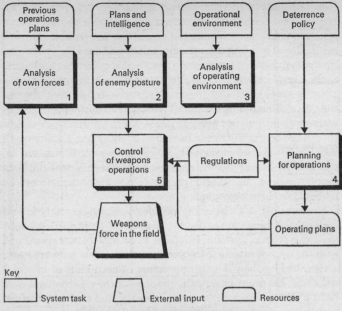

Figure 2 Analysis of a deterrent weapons control system
(fictitious example)

processing problem, data collection, data interpretation, data
manipulation, formulation of decision alternatives or restructuring of the original problem, decision making or problem solution,
data output.

Each block also represents a subtask that may require the use of
the facilities of the entire system or of only a very small part of the
system. The diagrams do not show the temporal order of task
performance. This may be developed by a 'high traffic' analysis
as described by H. H. Goode and R. E. Machol (1957). It would
be difficult to show time reference with the subject matter chosen
for the illustration because the processing activities seem to be
complexly interrelated in terms of recurring feedback.

This process of system task analysis is almost a type of 'boot-
strap' operation. There is evidence, however, in favor of the point
that different groups of people working on the analysis of the
same task and possessing comparable background information
produce similar designs. The particular arrangement of the blocks
in the diagram, the particular sequences, and the particular labels

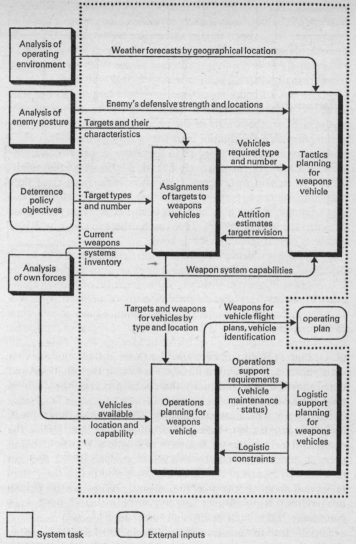

Figure 3 An example of subtask specification :
expanded view of the system task, 'planning for operations'

may differ. The essence of the similarity is contained in the specification of the data and their processing to accomplish the task objectives. This is encouraging and leads one to believe that such analytic diagramming may be as much a legitimate process in system analysis as are those processes whose rules are somewhat more formally defined, such as mathematical modeling.

At the risk of being repetitious, it should be pointed out that there are major advantages in employing this technique of task analysis. First, the results of the process, namely, the flow diagrams, furnish a substantial basis for unambiguous communication among the people involved in the design effort. Second, this method allows for the discovery of task relationships that might otherwise remain undetected. Only those who have tried flow diagramming can appreciate the fact that it is hard to close logical gaps. Finally, if such a technique is useful in practice, it takes us closer to the goal of being able to teach others to use it, which of course is one of the aims of this book.

One of the problems with the rational design process, as so far described, is that it tends to make the system look impersonal, as if it might well operate without the intervention of people. However, in successively more concrete levels of analysis, one sees that at various places human judgment would have to intervene to perform the function of data interpretation. Something similar to this process was described by M. S. Meeker, R. J. Rogers and G. H. Shure (1960) and employed by them in their analysis of a set of military command staff tasks. Essentially they analysed and diagrammed a task, indicating the decision points, the decision rules, and other data which enabled them to emerge from each decision point with a yes or no answer. Where the data so mapped, including regulations, were not adequate for determining the decision a priori, they indicated a need for a human decision process. Much more can be said about analysis of the decision processes by means of the block diagrams that define the system tasks, and about allocation of the decision prerogative to a person or computer. It is sufficient here to point out that all the human processing tasks, such as monitoring, comparing and assessing, eventually lead to decision making. The locus of decision making should be set primarily by user requirements rather than by some academic scheme for comparing and contrasting the strengths and weaknesses of men and computers.

From the analysis of user objectives and system tasks, we emerge with a set of general data requirements and definitions of subtasks of the type indicated in Figures 2 and 3. These diagrams are fictitious. They show how the task of planning for the use of deterrent weapons is analysed into subtasks, each with its necessary data types. These data essentially fall into two classes: those generated within a task and those coming into the task from another task or external agency. Figure 3 shows system subtasks, still at a fairly general level. In many instances the analysis of system tasks would have to proceed through several more levels before there would be enough detail to show all the specific data requirements for task performance. It is not possible to predict how many levels of analysis would be required for each task or for a system as a whole. Furthermore, not all tasks within the same system require the same number of analytic levels.

Let us assume, just for purposes of illustration, that Figure 3 represents a final level of analysis; this final level has been reached by answering the questions discussed under 'Block 5. System Task Requirements,' particularly the first four questions. The rational design process will then proceed by using the data so obtained to answer the questions indicated as decision hexagons in Figure 4.[8] These questions essentially are the same as those just used to develop task requirements. However, the emphasis here is on the last three questions. Each task is analysed separately, and the answers are compiled within the numbered rectangles as shown in Figure 4.

Once a system task is analysed in this way, the designer knows what data the system will have to obtain and process for that task. The required data flow can then be incorporated into the design document as follows:

1. By representing the data flow for each system task within the design format previously agreed upon by the user and the designer. This is analogous to 'single-thread analysis' as defined by Goode and Machol (1957, pp. 305–308). Each datum input to or created by the system is traced through the system until it exits or is destroyed or replaced. All intermediate processing and storage are shown and described.
2. By representing the data flow for *all* system tasks and building a composite. This is analogous to the first step in Goode and

8. If one looks carefully at the decision hexagons in Figure 4 and then at the attributes of the system elements, a high degree of similarity is perceived. Indeed it is up to the system designer himself to select a set of attributes so that they can be incorporated into questions such as those in Figure 4.

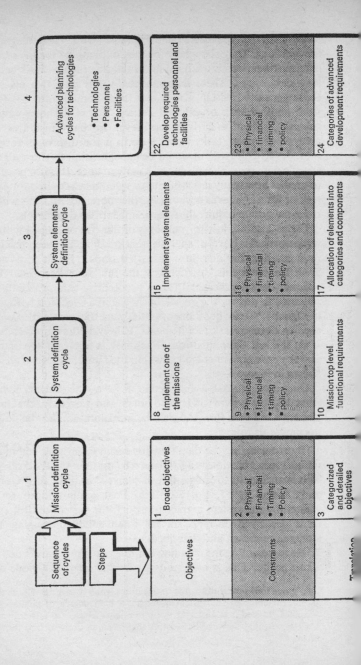

Sequence of cycles

| 1 Mission definition cycle | 2 System definition cycle | 3 System elements definition cycle | 4 Advanced planning cycles for technologies
• Technologies
• Personnel
• Facilities |

Steps

Objectives	1 Broad objectives	8 Implement one of the missions	15 Implement system elements	22 Develop required technologies personnel and facilities
Constraints	2 • Physical • Financial • Timing • Policy	9 • Physical • financial • timing • policy	16 • Physical • financial • timing • policy	23 • Physical • financial • timing • policy
Translation	3 Categorized and detailed objectives	10 Mission top level functional requirements	17 Allocation of elements into categories and components	24 Categories of advanced development requirements

Analysis	Possible missions and approaches with requirements	Possible detailed functions; system elements and their performance requirements	Possible design approaches for the components and their design requirements	Possible development tasks, schedules, decision points and funding approaches
Selection criteria	5 • Performance • cost/effectiveness • Timing • risk • Policy	12 • Performance • Cost/effectiveness • Timing • risk • Policy	19 • Performance • cost/effectiveness • Timing • risk • policy	26 • Performance • cost/effectiveness • Timing • risk • policy
Trade-off and synthesis	6 Selected missions and their broad requirements	13 Selected system elements and their performance requirements	20 Selected design approaches	27 Selected development tasks
Cycle output	7 Mission plan • Missions • Requirements • Timing	14 System definition system elements and their performance requirements	21 Advanced development requirements • Design approaches • Development requirements • Timing	28 Advanced development plans • Tasks • Funding • Schedules • Approach • Decision points

Figure 4 An example of system task analysis

Machol's 'high traffic analysis.' This is discussed below under 'Block 6 System Element Requirements.'

Figure 5 shows how the data flow resulting from the task analysis may be incorporated into a structured design document. Part of a system design document actually could be in the format illustrated. Each block in the diagram that contains a number in the lower right corner represents a system element, according to the key provided.[9]

The number of circles or ovals shows the blocks in the task analysis (of Figure 4) from which the data for this mapping are obtained. As a result of the task analysis, each block in the diagram would contain a complete specification of the relevant attributes of that element.

Block 6 System element requirements

After all tasks are analysed, using the single-thread procedure just discussed, the sum of the data within an element represents the total which that element is required to handle. The element must now be constructed so that it can perform its characteristic actions in the way called for by all the tasks. The element has to perform a certain number of its actions within given time and error limits. Also, the element is required to 'know' when it should start and stop operating and when it needs maintenance. Actually, mechanisms and procedures are required for activating the elements to accomplish the system tasks.

The specification of actual enabling mechanisms (equipment, computer programs, procedures) should not be stated until all the requirements for every element have been obtained and subjected to further analyses, including high traffic analysis.

Block 7 Integration of system elements

The requirements for system elements initially are stated in fulfilment of the set of ideal requirements that emerges from the analysis of system tasks. In the design process these ideal requirements are usually generated relatively independently of one another. They are also generated before very much work has been done on integrating them. This is probably as it should be, since it is more convenient to start from a situation with few interactions. However, it must be recognized that this interactionless state

9. It should be noted that the learning and retrieval elements are not included. At this point in the analysis we are interested only in the operational data flow.

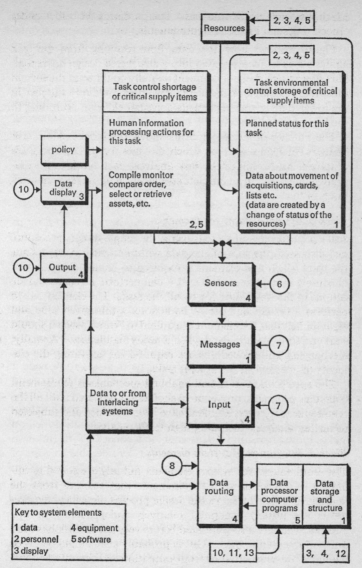

Figure 5 Representation of data flow for a system task

can only exist as a convenient assumption during the task analysis process. As soon as the designer attempts to integrate such independently derived requirements, he starts around the iterative loop indicated by the feedback from block 8 to block 5 in Figure 1.

A good example of the need for an integration process is furnished in the case of specifying the requirements for a display system in a military command post. In this example the following system elements have to be integrated: display, data, personnel and architecture.

1. *Display*. The number of display interfaces or display devices; the size (and other attributes) of the display data.

2. *Data*. The number of data to be displayed simultaneously.

3. *Personnel*. The number of people requiring access to a set of displays at the same time or over a period of time.

4. *Architecture*. The size and configuration of the physical area available; space requirements for passageways and equipment.

All of these elements must be specified. All of them are interrelated. All of them have to be integrated in such a fashion that the designer can make tradeoffs, knowing the ramifications of each trade. For instance, if he wishes to decrease the number of display devices without reducing the number of personnel requiring a fixed amount of simultaneous data, he then has to provide a display device for mass viewing and a room of a certain size and shape.

The extensiveness of the design integration effort is partly a function of the choice of design methodology. A reiterative method makes more provision for this analytic feedback effort. But even where a one-shot method is adopted, a considerable amount of time must be allocated for this process. Obviously, here is an effort which is a major consumer of time. Identification of it, as such, is a must for the adequate performance of the design project management.

Early examination and understanding of the more interactive attributes of the various system elements will help to prevent last-minute integration problems. Also, when these types of problems are recognized early, the design project management can take them into account. This can be done by partitioning the design effort in such a fashion as not to make different groups of people responsible for highly interactive system elements. One would not, for instance, partition off a display project from a data retrieval project.

References

ECHMAN, D. P. (ed.) (1961), *Systems Research and Design*, J. Wiley.

GOODE, H. H., and MACHOL, R. E. (1957), *System Engineering: An Introduction to the Design of Large-Scale Systems*, McGraw-Hill.

ROGERS, R. J., MEEKER, M. S., and SHURE, G. H. (1960), *Command Decision-Making* (classified), Systems Development Corporation, N10911.

14 Stanford L. Optner

Long-Range Plan for Systems

Stanford L. Optner, 'Long-range plan for systems', highlights of a report for the City of Los Angeles, 1968

Background

Between 1964 and 1967, the City of Los Angeles experienced rapid growth in the use of automatic data processing. The Board of Administration of the Data Service Bureau and the Office of the City Administrative Officer observed two striking phenomena during this period: first, a proliferation of systems installations in the Data Service Bureau; second, a proliferation of proposals for new systems from within and outside the City. The first phenomenon was an outgrowth of the 1963 data processing centralization ordinance which concentrated computer services in a single City computer facility. The second phenomenon was attributable to rising capabilities of systems analysts, in and out of the City, who saw new ways to use computers in municipal government.

During this period, the Data Service Bureau began the change-over from an IBM 1401 to the newer IBM System 360/30, a faster, more economical computer. At the outset, the Data Service Bureau operated the IBM 360/30 as if it were an IBM 1401, in order to avoid large reprogramming costs during this conversion period.

Problems of centralization, proliferation of systems and change-over to System 360 were accompanied by rapidly rising costs for personnel and equipment in the Data Service Bureau. From a City-wide total, exclusive of the Department of Water and Power, of 147 positions and a budget of $1,434,222 in 1962–63, the proposed budget for the fiscal year 1968–69 is for 257 positions and $4,039,955. City functions are now conducted on two IBM System 360/40 and one IBM System 360/30 computers, operating two to three shifts per day, six days per week.

With rapidly increasing use of data processing, the City Administrative Officer was frequently required to write reports recommending support or denial of requests of City departments for funds to study, design or implement new computer applications. In some instances, the facts were easily obtained and analyzed; in other instances, where the state of the art in municipal data processing was leap-frogging the experience of the City, there was less assurance that all the facts were available and the resulting conclusions properly drawn. As proposals became more elaborate, the financial consequences of inappropriate decisions were disproportionately greater.

The Data Service Bureau generated a City-wide concept of systems in 1965 called LAMIS (Los Angeles Municipal Information System). The principle of a unified City system has been generally accepted since that date. However, because of the administrative, technical and financial questions raised by potential future uses of computer technology, the City Administrative Officer recommended that data processing consultants specializing in municipal information systems be retained to review and update LAMIS and to define a systems development plan for City departments. The Board of Administration studied the need for a Long Range Plan for Systems, approving the proposed study in June, 1967. An inventory of possible computer applications was assembled by the City Administrative Officer to guide Stanford L. Optner and Associates, who were selected to execute the study.

Investigation

A number of departments indicated interest in the project to develop a Long Range Plan for Systems. Several submitted requests for data processing service in applications areas which cut across departmental lines, involving activities common to many City departments. Others took a less City-wide view and confined requests to problems within their own departments. The Technical Report of this study deals specifically with the inventory of potential data processing applications which were assembled as a guide for the investigators.

Some applications were added to or deleted from consideration based upon the following criteria:

1. Does the computer bring an important added capability to the problem?

2. Do the system content, file size, record size or frequency of processing warrant the use of automated techniques? Is the speed of entering, processing or reporting data currently satisfactory?

3. Does data analysis deal with the problem completely and adequately?

4. Do incremental start-up and annual operating costs justify conversion to the computer?

5. Do the service benefits outweigh the start-up and annual operating costs?

Other possible computer applications were added to those proposed by departments for the following reasons:

1. Although some departments have had considerable experience with data processing, others were unable to identify likely tasks because of a lack of understanding of what computers can do.

2. Some departments would prefer to continue their current method of operation and will do so unless they are requested to take advantage of the availability of computer technology.

3. No single department can be expected to design a technique for dealing with a problem which is not entirely within its own jurisdiction.

Some departments with many years of tabulating or data processing experience submitted a large amount of data consisting of existing and future potential applications. The vestiges of tabulating-oriented systems were still apparent in the course of this study. Some departments have made great use of Data Service Bureau equipment and/or services whereas others have had little or no exposure to computer technology. The investigators confined their work to those applications which met the above criteria and also appeared to have some future impact on Data Service Operations workload.

The current position of the Data Service Bureau with respect to computer and peripheral equipment, personnel and facilities was also reviewed. This dynamic organization is in a state of transition. It is maintaining a high level of day-to-day service and simultaneously changing its way of executing workload. Such changes are typified by new mass memory equipment, new programming concepts to reduce costs of file maintenance, new investigations of programming techniques to minimize the cost of special purpose reports and new large-scale systems designs. In this report, two equipment configurations are suggested for consideration in the next five years.

We briefly investigated the possible use of generalized programming techniques, an innovation in computer software which promises to reduce the one time costs of file maintenance, information retrieval and special-purpose programming. Our brief report to the Data Service Bureau suggests that the use of proprietary programs for this purpose is not in the best interest of the City. IBM expects to release an initial version of a generalized programming system called GIS in the fall of 1968. We have suggested that GIS, when perfected, will do as much or more than other similar programs at no added cost to the City. We, therefore, have recommended that the Data Service Bureau work closely with IBM in seeking to introduce these new programming concepts.

A number of new criteria were evolved to assist the City Administrative Officer Wage and Salary Administration division in its personnel analyses of Data Service Bureau staff. Quantitative and qualitative factors, useful in reviewing salaries of technical and professional positions for systems, programming and computer operations have already been submitted.

Facilities of the Data Service Bureau are currently scheduled for transfer to City Hall East in 1970–71. Assistance in advance planning of these facilities centred upon industrial engineering considerations such as physical space layout and materials handling. Specific recommendations have already been made in the following substantive areas:

1. Space utilization and location of stairwells.

2. Use of conveyors to stack printing supplies in the computer area, delivering from a ground floor or basement warehouse.

3. Use of conveyors to deliver unprinted paper and forms to the computer printer, removing same to a data control area after processing

4. Use of conveyors to deliver post-computer processed materials to mail room and shipping dock.

5. Use of conveyors between City Hall and City Hall East via the planned walkway above Main Street.

Investigators explored the concept of a citizen information center on the ground floor of City Hall East. There has been considerable interest in the past in the use of such a facility employing remote input/output devices to interrogate the computer. There has also been interest in a central license and permit issuing area within City Hall East. Neither of these projects are as yet in a sufficiently advanced state of development to be successfully evaluated. They would, obviously, create added computer operations requirements.

Two departments were found to have exceptionally advanced concepts of data processing/information systems which have an important, but as yet unevaluated, bearing upon the future of computer technology on the City. The Fire Department is in the initial stages of formulating the design concepts for an automated fire dispatch system; the Police Department is in a somewhat similar status with respect to a proposed command and control system. Neither of these projects were at a sufficiently advanced state during the investigation period to be analysed and incorporated in the Technical Report.

A number of possible revenue-producing measures were under discussion by the City Council at the time this report was completed. Only one of these, a proposed sanitation charge, has been studied to determine the City's alternative means of collecting the revenue. Should any of these measures be passed, or should State witholding become a law, there would be an additional, unevaluated data processing impact upon the City. There are findings associated with each department and each systems project. Listed below are only those findings which affect all City departments and synthesize the conclusions growing out of the Long Range Plan for Systems investigation.

The cream has been skimmed from the top: The obvious and most easily justified computer applications are scheduled for early conversion, are in the development stage or are already in operation. The first wave of benefits from automatic data processing have been realized. They are in two categories:

City departments have accepted the principle of centralization of computer service: centralization has eliminated the need for at least ten departmental computers and the associated staffs. Further savings from centralization will be realized when all keypunch and tabulating services of departments are consolidated into the Data Service Bureau. The savings from centralization of equipment and personnel currently approximate $2·5 million annually. Savings in 1970–71 may reach $3·4 million annually as shown in Table 1.

Major programs with public service, management, and dollar benefits are in process: the Data Service Bureau is serving all

Table 1 Estimate of costs conserved by centralizing data processing services in the city of Los Angeles

	1968-69		Forecasted costs 1970-71		1973-74	
	Actual budget	*Estimated budget if decentralized*	*Centralized*	*If decentralized*	*Centralized*	*If decentralized*
Computing equipment	$ 1,016,766	$ 2,038,524	$ 1,482,450	$ 2,655,540	$ 1,939,548	$ 3,233,544
Keypunch	38,616	55,040	40,029	71,700	24,244	40,419
Tabulating	100,974	144,735	51,889	92,944	38,791	64,670
Basic equipment rental	1,156,356	2,238,299	1,574,368	2,820,184	2,002,583	3,338,633
Other equipment	333,875	407,370	393,615	513,273	500,696	607,631
Total equipment rental	1,490,231	2,645,669	1,967,983	3,333,457	2,503,279	3,946,264
Added expense	480,138	330,709	454,253	416,682	450,626	493,283
Sub-total	1,970,369	2,976,378	2,422,236	3,750,139	2,953,905	4,439,547
Salaries	2,058,272	3,538,913	2,376,378	4,458,915	3,258,547	5,278,621
Furniture/fixtures	11,314	19,545	14,094	24,627	18,638	29,155
Total	$ 4,039,955	$ 6,534,836	$ 4,812,708	$ 8,233,681	$ 6,231,090	$ 9,747,323
Costs conserved		$ 2,494,881		$ 3,420,973		$ 3,516,233
Budget total[1]	$420,418,800	$422,913,700	$468,821,000	$472,241,883	$709,579,000	$713,195,000
Percent of budget total	·96	1·55	1·03	1·74	·88	1·37

[1] Forecasted costs are based upon data processing capabilities proportional to workload as evidenced from EXHIBIT G of the proposed fiscal year 1968–69 budget. Data are for appropriations under the control of the Mayor and Council.

departments of the City with a range of applications from library science to refuse collection, from fund-appropriation accounting to traffic accident reporting. There are over 1,100 active computer programs which cycle through the computer center. Los Angeles is pioneering in some applications areas, and has served as a model to other jurisdictions. A few computer applications are returning substantial dollars of savings to the City to offset original start-up and annual operating costs. More dollars may be saved, depending upon the selection of areas to be automated and the rigor with which potential savings are pursued.

The City must change its way of doing business to obtain a second round of benefits from computer technology: the economic and operational advantages yet to be realized are the most valuable which data processing may render. But they cannot be obtained by working around existing systems and organizational practices which have evolved over the last fifty years. The second round of benefits will result from new activities which must be launched by the City:

The City should employ more of the available Data Service Bureau resources in the design and implementation of large-scale systems which cross departmental lines: data processing problems common to many City departments demand a large-scale commitment of man-power. They should not be studied on a piece-meal basis during systems analysis and design. The piece-by-piece approach to municipal problem solving perpetuates the independence of departmental systems designs, and does not facilitate the achievement of City-wide systems objectives illustrated in Figure 1.

Demonstration projects conducted in the fiscal years 1966–67 and 1967–68 have revealed the interdependence of systems and have proven the practicality of integrating the input/output relationships of systems. The concept of continuous stream processing for operations reporting; payroll accounting, cost accounting, personnel reporting, performance budgeting and budget formulation is no longer a theoretical, but a practical objective. What is needed now are systems projects of large scope aimed at realizing a new generation of data processing benefits.

The City should provide more technical support to departments with information systems requirements: the systems essentially contained within a single department are integral to the fulfillment of day-to-day City operations. Despite the need, only one

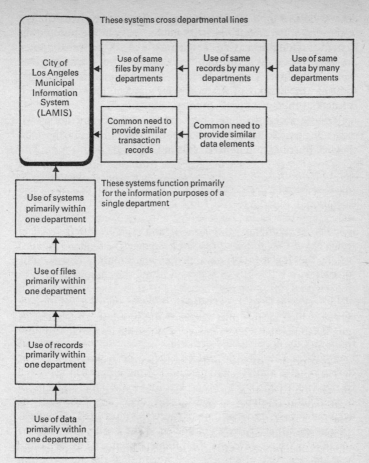

Figure 1 Integrated systems design will facilitate the achievement of city-wide and departmental objectives

department is in the process of installing an automated management information system.

Existing departmental methods of providing management control become more complex as a result of increased workload. Manual systems tend to build up new requirements for personnel and levels of organization as workload expands. Data Service Bureau efforts should be invested in the design and implementation of information systems which restructure operating methods through the use of computer technology.

The City should increase its use of specialized electronic and electro-mechanical equipment: the use of special purpose equipment to acquire operations data at the point of the transaction in machine-processable form holds the potential of improving cost and systems performance. Special purpose equipment will make possible the design of systems which accent improvements in speed and effectiveness of communication.

Remote access to mass, high-speed electronic memories will make it possible to communicate with central files in the Data Service Bureau from departments in City Hall and from outlying City locations. The benefits accruing from these innovations will be primarily functional. Departments will have shorter reaction times, will make better informed decisions and will have improved management control.

The new generation of systems designs should bring selected, non-confidential City files and records continuously on-line to qualified users: the logistics of data processing should change. Why operate at a millionth of a second inside the computer, or print reports at 1100 lines per minute, and then take four hours to deliver a report to its users? What is the real value of the automated system which swings into action at the speed of light, after two months of manual system delays in preparing input data for computer processing?

Appropriate City files and records should be updated when the transaction is made, with the speed and efficiency of an airline reservation. Departments which need City data immediately must have as rapid access to time-current files and records, as the user of a dial telephone. Growth should be accompanied by change, to modernize the way the City conducts its business. The equipment components to bring about on-line real time processing are shown in Figure 2.

Data processing systems should be viewed as a means of communicating information: data acquisition must by-pass conventional modes of data conversion which are proportional to the volume of transactions. Remote data conversion and transmission to the Data Service Bureau should replace, where possible, labor-consuming steps which are entailed in key-punch, batch processing and serial file maintenance.

Communication systems must be most heavily employed in departments with decentralized field operations and offices located outside of the Civic Center. Communications systems for the

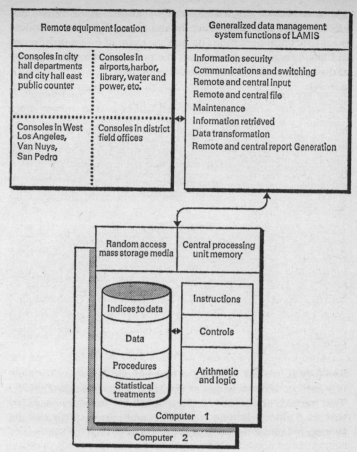

Remote equipment location		Generalized data management system functions of LAMIS
Consoles in city hall departments and city hall east public counter	Consoles in airports, harbor, library, water and power, etc.	Information security Communications and switching Remote and central input Remote and central file Maintenance Information retrieved Data transformation Remote and central report Generation
Consoles in West Los Angeles, Van Nuys, San Pedro	Consoles in district field offices	

Random access mass storage media | Central processing unit memory

Indices to data

Data

Procedures

Statistical treatments

Instructions

Controls

Arithmetic and logic

Computer 1

Computer 2

Figure 2 Equipment components to bring about on-line real-time computer processing

Departments of Water and Power, Harbor and Airports, and for branch city halls are important in this regard.

Implementation benefits

Growth in the size of City government and the complexity of conducting its work should be paralleled by important changes in business procedures and methods. Current methods perpetuate organization practices which were developed when Los Angeles was a smaller city in a simpler environment. The City has not gone far enough in upgrading the technique of providing services which are dependent upon manual and machine data processing. If the City addresses and solves the problems brought about by growth, these are the benefits to be obtained:

Faster access to files, records and data means that City business will move at a faster rate: city employees will be able to answer more questions within the same time period. Decisions will be reached at a faster rate than now possible under cyclical data processing. Today, the report user is delayed by the time it takes to obtain the most current information before taking an informed action. Immediate access to records and automated analysis of data will speed decision processes. Information will move through City departments faster.

Restructured City systems will bring about improved use of Data Service Bureau resources: day-to-day operations will be characterized by use of multi-processing, remote random access and source data automation in an on-line, real-time, man-machine environment. These conditions will bring about more intensive use of machine facilities. More powerful computer equipment with more high speed peripheral mass storage devices will be required. Increased costs of equipment to provide these capabilities will be less than the alternative course of inaction.

The City should not continue to pyramid one system on the next, in a haphazard development of independently implemented efforts: This trend can only lead to the requirement for more computer time, and ultimately more computers and personnel than would be required under the proposed Long Range Plan for Systems.

Future increases in workload will not require proportional increases in City staff: fewer manual tasks will translate into a need for fewer personnel to handle future quantities of transactions. Fewer manual operations will be required to execute system functions. The position of the individual with respect to the system will change. Individuals will have fewer clerical tasks to perform; their functions will be more judgmental in nature. Tasks will proceed with more directness and less delay from the startup to the conclusion of the system functions.

There will be fewer redundant operations: individuals will not re-enter transcribed data, as they must currently do, in order to execute required systems operations. The duplication of operations in the acquisition and conversion of data into machine-processible language will be eliminated where economical. The emphasis on source data automation will make it possible to direct transactions to the proper file without the need for multiple personnel handlings.

Increased use of remote equipment will reduce the elapsed time to maintain central files: the increased use of remote equipment in source data automation will make possible the on-line maintenance of master files. The use of remote file maintenance and information retrieval will help to stabilize the costs of systems operation. Figure 3 illustrates the data processing interactions of on-line inquiry to City files.

Remote file maintenance will replace more costly data processing techniques which currently prevail: batch processing and cyclical handling of data will be eliminated where possible. Serial file maintenance on magnetic tape, which may require many computer hours, will continue to be employed only in systems where it remains economical.

There will be increased flexibility in the format and content of reports: users of special-purpose information will not have to rely on fixed format reports which contain general-purpose information. They will be able to extract records or data, operate on them, transform them in a variety of statistical treatments and display the solution of a specific problem in visual or printed form.

System users entitled to enter the file will command specific data in the file instead of the voluminous pages of a report. Much of what is displayed in elaborate computer print-outs is never read by the report user. Computer technology now makes it possible to specialize reports to the information germane to a particular problem under study. Information users will spend less time in data review, data handling, redundant page scanning, surplus data acquisition and will function with improved effectiveness.

System users will have access to central files for data originated in other departments: some departments need access to data contained in the records of other departments. Some of these are relatively inaccessible for today they can only be acquired by manual transcription. Improved access to the City's information base will result in more effective operations of individual

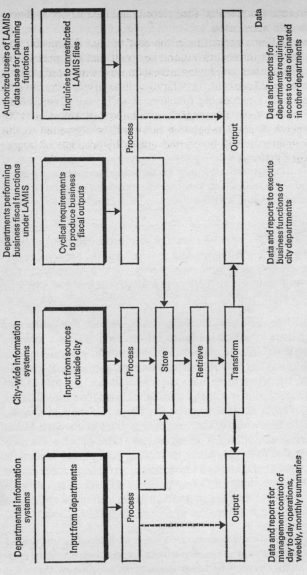

Figure 3 Data processing interactions during on-line real-time inquiry to Los Angeles Municipal Information System (Lamis) files

Departmental information systems

Input from departments

Process

Output

Data and reports for management control of day to day operations, weekly, monthly summaries

City-wide information systems

Input from sources outside city

Process

Store

Retrieve

Transform

Output

Data and reports to execute business functions of city departments

Departments performing business fiscal functions under LAMIS

Cyclical requirements to produce business fiscal outputs

Process

Authorized users of LAMIS data base for planning functions

Inquiries to unrestricted LAMIS files

Output

Data

Data and reports for departments requiring access to data originated in other departments

departments. Confidential files, records and data will be available only to authorized users.

There will be a reduction in the cost of data conversion and tabulating over the next two years: key-punching will yield to use of data converters employing magnetic tape with possible cost reductions of 25 per cent. Tabulating will be largely eliminated as a primary data processing function. It will become peripheral, aimed at the most marginal low volume operations of City departments. Current tabulating tasks will be converted to the card computer or will be carried out as by-products of larger, automated systems.

15 Donald F. Heany

Requirements, Concepts and Approval

Excerpts from D. F. Heany, *Development of Information Systems*, The Ronald Press Company, 1968, pp. 45–71

It is convenient to visualize system development as a process with eight discrete steps:

1. Establishing or refining an information requirement.
2. Developing gross system concepts.
3. Obtaining approval to detail a particular gross system concept.
4. Preparing detailed system specifications.
5. Testing.
6. Implementing.
7. Documenting.
8. Evaluating the system and the process followed in putting the system in place (Figure 1).

These steps in system development often overlap. It is not always possible to finish one step before going on to the next. Furthermore, the process is iterative; designers are often forced to recycle.

Step 1 Establishing the information requirement

Work in Step 1 in system development is focused on developing, refining, and/or reconciling one or more information requirements. All parties contributing to the system must understand and agree to what is wanted, who wants it, when, where, and why. These are the classic questions. Unfortunately, they constitute a deceptively simple formulation of the work to be done in Step 1. When a designer leaves behind unifunctional, routine systems, he discovers that information requirements and operating work are inextricably intertwined. To get at the requirements he must first understand in great detail the work being done in the operating organization under study, and the way its key people view that work.

End result of Step 1

Step 1 should end with three outputs:

1. A statement of the system objective (or objectives).
2. A list of those environmental features most likely to affect the system.
3. A list of the restrictions that bear upon the scope of the system or on the way the designer proceeds.

System objective: the term 'system objective' is shorthand for the business benefits sought from the proposed system study.

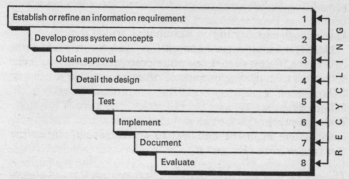

Figure 1 The generic steps in the development of a single information system

System objectives are specific targets indicating how the designer's efforts will support operating work. If the real purpose of information-systems work is to improve the effectiveness of the business, then system objectives will be expressed in terms of what operating people will be able to do once their stated information requirements have been satisfied.

Constraints: the final element of a refined information requirement concerns restrictions on the designer's freedom of action explicitly or implicitly stated by users of the system, by their managers, or by the designer himself. For example, the format of data, the way they are coded, and the length of time they are to be retained may be determined by law, by internal policy, by a contractual agreement with a customer, or by some trade-association agreement.

Security rules may be imposed on information systems that

Donald F. Heany 275

process data for military procurement programs. Personnel information systems must be designed to preserve the privacy of individuals. There may also have to be an audit trail for any dollar figures generated by the system.

Many constraints derive from resource limitations. How many designers and programmers are available for assignment to this one system study? When is the output due? How many dollars are available for data collection? Is the designer free to talk to all experts in the area served or are some too busy? The limiting resources may be outside the company. For example, a designer of a corporate employee savings plan discovered that he had to provide for the disbursement of $50 United States Savings Bonds because the Federal Reserve Bank's computer facility did not have enough capacity to run $25 bonds.

Not all constraints are imposed. The designer may add some of his own. He may scale down the posed requirement in order to be able to utilize the outputs of other information systems or to exploit some characteristics of in-place processing or communication equipment. He may elect to restrict the scope of investigation or the type of data to be gathered. For example, a new order entry system might be designed to accommodate only orders from major customers, and then orders for items in the sales catalog that accounted for 80 per cent of last year's sales. If a business has five product lines, the designer might design an information system for one line. Next year he can test his assumption that this system can be readily adapted to other product lines.

Step 2 Development of gross system concepts
Examples of system concepts

The following three examples of system concepts have been chosen to dramatize the importance of approaching information requirements without preconceived ideas.

Retrieve or regenerate engineering data? Businesses that offer customers hundreds or thousands of variations of basic designs must contend with the problem of storing, retrieving, and updating engineering drawings. At times, the volume of drawings becomes so large that it is simpler for engineers working on a new order to make a new drawing and assign it a drawing number rather than to search old drawings for one that might apply. Consequently, several drawing numbers often exist for the same drawing.

Systems designers have proposed (and some have implemented) a gross system concept which promises to reduce this storage, retrieval, and maintenance problem. Instead of looking at engineering drawings as physical objects to be stored – like books in a library – the idea is to provide engineers with the capability of automatically regenerating all the drawings and manufacturing planning paper needed to fill a customer's order *as it arrives*. This concept requires the business to store and maintain certain basic engineering data plus the routines for converting these data into the required engineering and manufacturing information on demand. Each order is treated as if it were unique. No effort is made to compare it with other orders. None of the output generated for a customer's order is saved.

Contrasting definitions of manufacturing's job: a firm manufacturing products for public and private utilities was anxious to reduce the time required to make and ship its products. Automation of work stations in Manufacturing had already been pressed to the limit. A suggestion to communicate and process orders from customers and district offices electronically rather than manually was under consideration. This proposal implied that the existing method of ordering, making, and shipping products was fundamentally sound but that the efficiency of certain information subsystems could be improved.

After studying this proposal, the chief executive of this company elected to try a different approach which affected the information systems in Manufacturing and in Engineering. It called for Manufacturing to plan production on the basis of subassemblies and parts common to many models. Previously, Manufacturing had built its schedules around end items shown on Marketing's short-term forecasts.

This change greatly simplified Manufacturing's scheduling and dispatching problems. Even though orders for individual models fluctuated in an erratic fashion, the volume of subassemblies and parts required to sustain shipments was remarkably stable. Operating people had always appreciated that their products shared components. What they did not appreciate was the extent of this sharing and how to take advantage of it. They had had no way of stabilizing production and simultaneously helping Marketing render the service customers were demanding.

Empirical approach: some designers begin Step 2 by amassing as much information as possible about the in-place system to be replaced or improved. They justify their approach on four grounds. First, there is always a chance that an inexpensive modification in existing information systems will satisfy the objective without a full-scale systems study. Should this examination indicate that in-place information systems are truly inadequate, these designers still feel that they are rewarded handsomely for their efforts. They acquire knowledge of what they are about to change, and this will be most important when the time comes to implement the proposed system. Regardless of their inadequacies, in-place systems are understood by the operating people served. They have these systems calibrated. They have learned when to depend on them and how to compensate for their inadequacies. Naturally, users will hesitate to accept a new system from a designer who displays no understanding or appreciation of the good features of their old system.

Logical approach: other equally talented designers try not to get too close to in-place systems lest this compromise their objectivity. To identify with operating people or with the designers of the current system might, in their judgment, channel their thinking. Instead, these designers concentrate on logical methods of satisfying the posed requirement. They try to be uninhibited in Step 2, to consider all conceivable ways of deriving the information. In this, they argue, ignorance of details of how things are actually done is a blessing in disguise. They feel free to experiment with novel ideas which an individual more knowledgeable in the ways of the business might dismiss as impractical.

Step 3 Obtaining approval

The third step in system development is in part a progress report and in part a request for approval to proceed with development in a particular manner.

Purposes

The designer's primary purpose is to get operating managers to authorize the completion of system development. Their approval may or may not include a commitment of additional resources.

The designer has certain secondary objectives as well. He is anxious to build some understanding of and enthusiasm for this proposed system. He does not want to plunge deeply into detailing and implementation on a mere act of faith in his professional competence. Managerial understanding is a much more solid basis for later steps in system development. From this point on, it will become more and more difficult to adapt to changes in direction. It is best to learn about any managerial reservations immediately.

Differences in design philosophies are best identified and defended in an open session with the users, who, in the final analysis, foot the bill. Such differences rightfully come under management review in Step 3.

The operating managers who listen to this status report and proposal for continuing design work have their own distinct objectives for Step 3. First, they are eager to determine how well the systems designer has understood their version of the information requirements. Has his effort to refine their initial statement endangered or distorted the objective? Has he introduced simplifications that will compromise the usefulness of information resulting from the proposed system? Does he have a good grasp of the business needs that led to this information-systems study in the first place?

Part Three
A Look Forward

Writing for the management scientists, Churchman reviews the part of the systems analysis that deals with choice of objectives, boundary definition ('environment'), resources and components of systems. He also discusses the management function in the systems approach, and raises the interesting question of whether the systems analyst lies inside or outside the system. The control function of management, in directing the systems analyst, emerges as a critical factor in the success of a systems analysis.

The authors of Reading 17 have been among the most active proponents of the use of quantitative method in systems analysis. Their example illustrates its use in organization theory, and suggests that there can be a general systems orientation to problems of social organizations. This approach to problem solving is among the most useful and powerful outgrowths of systems analysis and illustrates, perhaps better than any other article, the new role of management science in systems analysis.

The Mitchel material is of interest because it documents carefully the rise of the new systems analysis discipline, and in addition contains a discussion of the idea of feedback and its impact upon organization. This material is also of interest because it was pointed at government, and not at the military or the industrial establishment. Systems analysis is now playing an increasing role in the design of governmental systems. This article points to the future because of the probability that systems analysis techniques may influence the reorganization of State and local government functions and operations.

The brief piece by Simon Ramo indicates a rising use of systems-oriented techniques to deal with problems of big cities, environment and ecology, and other social issues of our time. Not with the view that they are a panacea . . . none of these writers have struck that note . . . but because they have the capability of providing the framework in which man may assemble all his resources to deal with the problems of living.

16 C. West Churchman

Systems

Excerpt from C. West Churchman, *The Systems Approach*, Dell
Publishing Company, Inc., 1968, pp. 28–47

There is a story often told in logic texts about a group of blind
men who are assigned the task of describing an elephant. Because
each blind man was located at a different part of the body, a
horrendous argument arose in which each claimed to have a
complete understanding of the total elephantine system.

What is interesting about this story is not so much the fate of
the blind men but the magnificent role that the teller had given
himself – namely, the ability to see the whole elephant and consequently observe the ridiculous behavior of the blind systems
describers. The story is in fact a piece of arrogance. It assumes
that a very logically astute wise man can always get on top of a
situation, so to speak, and look at the foolishness of people who
are incapable of seeing the whole. This piece of arrogance is
what I called 'management science' in the last chapter [not included here].

The arrogance cannot be allowed to remain unchallenged.
Only if we could be sure that the objectives of the management
scientist were pure and really in line with those of the total system,
and only if we could be sure that he had the observational powers
comparable to those of the observer of the blind men could we
feel that the scientist had the ability to see the whole.

But in the spirit of the debate, let's allow the management
scientist to describe how he climbs to the vantage point from
which he can view the whole system. His method is one of defining
carefully what he's talking about. He begins with the term
'system'. Although, he says, the word 'system' has been defined
in many ways, all definers will agree that a system is a set of parts
coordinated to accomplish a set of goals. An animal, for example,
is a system, a marvellously contrived one, with many different
parts which contribute in various ways to the sustaining of its
life, to its reproductive pattern, and to its play.

In order to make this definition more precise and also more

useful, we have to say what we mean by 'parts' and their co-ordination. Specifically, the management scientist's aim is to spell out in detail what the whole system is, the environment in which it lives, what its objective is, and how this is supported by the activities of the parts.

To develop this thinking further, we shall have to lay out a series of thinking steps, much as any manual of logic or rhetoric attempts to do. The reader should bear in mind, however, that these steps are by no means steps that must be taken in sequence. Rather, as one proceeds in thinking about the system, in all likelihood it will be necessary to re-examine the thoughts one has already had in some previous steps. Logic is essentially a process of checking and rechecking one's reasoning.

With this in mind, we can outline five basic considerations that the scientist believes must be kept in mind when thinking about the meaning of a system:

1. The total system objectives and, more specifically, the performance measures of the whole system.
2. The system's environment: the fixed constraints.
3. The resources of the system.
4. The components of the system, their activities, goals and measures of performance.
5. The management of the system.

It goes without saying that there are other ways of thinking about systems, but this list is both minimal and informative.

The objectives of the overall system are a logical place to begin, because, as we have seen, so many mistakes may be made in subsequent thinking about the system once one has ignored the true objectives of the whole.

Now the scientist's test of the objective of a system is the determination of whether the system will knowingly sacrifice other goals in order to attain the objective. If a person says that his real objective in life is public service and yet occasionally he seems quite willing to spend time in private service in order to maximize his income, then the scientist would say that his *stated* objective is not his *real* objective. He has been willing to sacrifice his stated objective at some time in order to attain some other goal.

In order to clarify the matter, the scientist needs to move from the vague statement of objectives to some precise and specific measures of performance of the over-all system. The measure of performance of the system is a score, so to speak, that tells us how well the system is doing. The higher the score, the better the performance. A student in class often comes to think of his objective as the attaining of as high a grade as possible. In this case the measure of performance becomes quite clear, and it is interesting to many a teacher to note that students seek to attain a high grade even at the sacrifice of a real understanding of the content of the course. They seek the high grade because they believe that high grades will lead to scholarships and other opportunities in the future. Their *stated* purpose is to learn, but their *real* measure of performance is the grade.

But to the thorough-thinking management scientist, this objection is not a serious one. In thinking about systems, he replies, we must move from what is often the real objective of the system managers to wider considerations. We may in fact have to begin to consider how to put the cost of accidents and ugliness into our measures. Intangible as these may be, he says, we shall see that the measurement of them is really not so difficult as might appear at first sight. In fact, there are some excellently worked-out cases in which highway engineers, as well as the designers of aircraft, have developed measures of the cost of an accident, in terms of the lost capability of the individual in earning income throughout the rest of his life. To the humanist, this may seem a very crass way of putting a number on the loss of a limb or a head, but to the management scientist it is the only practical way in which we can think about the so-called intangible aspects of the systems. In other words, he says, if we want to *think* about how loss of life or happiness or beauty is related to system performance, we can't simply say that these are so elusive that they cannot be defined, because by saying this we mean that we don't want to think about them at all. In order to think about them satisfactorily, we are going to have to be explicit and make our stand on the way in which these aspects of systems enter into measures of system performance.

Thus in the determination of a measure of performance, the scientist will seek to find as many relevant consequences of the system activities as he can. Admittedly, he too will make mistakes and will have to revise his opinion in the light of further evidence. But his persistence and alertness, and his intent to be as objective as possible, will enable him, he believes, to minimize his errors.

Supposing that some success has been attained in determining the system objective ('measure of performance'), the next aspect of the system the management scientist considers is its environment. The environment of the system is what lies 'outside' of the system. This also is no easy matter to determine. When we look at an automobile we can make a first stab at estimating what's inside the automobile and what's outside of it. We feel like saying that what lies beyond the paint job is in the environment of the automobile. But is this correct? Is it correct to say, for example, that what lies beyond the paint job of a factory is necessarily outside of the factory as a system? The factory may have agents in all parts of the country who are purchasing raw materials or selling its products. These are surely 'part' of the total system of the factory, and yet they are not usually within its walls. In a more subtle case, the managers of the factory may belong to various political organizations through which they are capable of exerting various kinds of political pressures. Their political activities in this case certainly 'belong' to the system, although again they hardly take place within the 'shell' of the system. And, returning to the automobile and considering what it is used for, we can doubt whether its paint is the real boundary of its system.

Hence the scientist has to have a way of thinking about the environment of a system that is richer and more subtle than a mere looking for boundaries. He does this by noting that, when we say that something lies 'outside' the system, we mean that the system can do relatively little about its characteristics or its behavior. Environment, in effect, makes up the things and people that are 'fixed' or 'given', from the system's point of view. For example, if a system operates under a fixed budget that is given to it by some higher agency and the budget cannot be changed by any activities of the system, then we would have to say that the budgetary constraints are in the environment of the system. But if by some organizational change the system could influence the budget, then some of the budgetary process would belong inside the system.

Not only is the environment something that is outside the system's control, but it is also something that determines in part how the system performs. Thus, if the system is operating in a very cold climate so that its equipment must be designed to withstand various kinds of severe temperature change, then we would say that temperature changes are in the environment, because these dictate the given possibilities of the system performance and yet the system can do nothing about the temperature changes.

One of the most important aspects of the environment of the system is the 'requirement schedule'. In the case of an industrial firm this consists of the sales demands. Of course in some sense the firm can do something about the demands by means of advertising, pricing, and the like. But to the extent that the demand for the firm's products is, so to speak, determined by individual people outside who are the customers of the firm, then the demand lies in the environment of the system, because it is a 'given' and because its nature influences system performance.

Here again we get some insight into the character of the management scientist. The environment is not the air we breathe, or the social group we belong to, or the house we live in, no matter how much these may seem to be outside us. In each case, we must ask, 'Can I do anything about it?' and 'Does it matter relative to my objectives?' If the answer to the first question is 'No' but to the second is 'Yes', then 'it' is in the environment.

Next we turn to a consideration of the resources of the system. These are *inside* the system. They are the means that the system uses to do its jobs. Typically when we turn to the measurement of resources we do so in terms of money, of man hours, and of equipment. Resources, as opposed to the environment, are the things the system can change and use to its own advantage. The system can decide which of its men shall work on which jobs, or how its money shall be spent on various activities, or what the time limits will be on various kinds of activity.

Within many systems a very careful analysis is made of resources. The traditional company balance sheet in effect is a listing of the various kinds of resources that a firm has available, especially when these resources can be translated into money: buildings, equipment, accounts receivable, cash, etc. But the management scientist has concluded that the traditional balance sheet leaves out many of the important resources of a firm. It does not give a detailed account of the type of personnel that the firm has available in terms of their educational background and personal capabilities, for example. Something like 'good will', which is surely a resource, is often represented by a fictitious number on the balance sheet.

For the management scientist, the systems approach entails the construction of 'management information systems' that will record the relevant information for decision-making purposes and specifically will tell the richest story about the use of resources, including lost opportunities.

There is another aspect of resource determination that is quite important in an age of expanding technology: it is quite essential that firms and government agencies pay special attention to technological advances that may be able to increase their resources enormously. In looking and thinking about a system, the management scientist pays attention not only to existing resources but also to the manner in which resources can be increased, that is, to the manner in which the systems resources can be used to create better resources in the future, by means of research and development in the case of hardware types of equipment, or by training and education of personnel, and by various kinds of political activities which will increase the budget and investment potential. In fact, for many systems a component that deals with the increase of resources may be the most important component of the system.

Resources are the general reservoir out of which the specific actions of the system can be shaped. The specific actions are taken by the components, or parts, or subsystems (all these terms being used interchangeably in management science). Components is the fourth item in the 'thinking' list. Here again, says our scientist, our thinking is apt to be blurred by tradition. Organizations are often divided into departments, divisions, offices, and groups of men, but careful examination shows that these are not the real *components* of the system even though they carry labels that seem to indicate that they are. For example, in industrial firms a department may be labeled 'production'; this should lead us to think that only within this component can one find the manufacture of products. Another department will be labeled 'marketing'; one should therefore believe that only in that department would one find the activities dealing with distribution and sales of products. And yet in many firms the distribution function must be conceived as part of the production component simply because it would be quite impossible to determine how the distribution of products should occur independent of the way in which the products are made. And perhaps the production department has a great deal to do with the manner in which products are sold simply because production must deal in many cases directly with the customer in satisfying his orders. If the customer is badly disappointed, then the activities of the production department may decrease sales.

It is for this reason that in thinking about systems the management scientist ignores the traditional lines of division and turns instead to the basic 'missions' or 'jobs' or 'activities', all of these labels being used to describe the same kind of thing, namely, the rational breakdown of the tasks the system must perform. Thus in the case of a city or a state, the basic missions may be defined in terms of health, education, recreation, and the like. If they are so defined, the scientist sees that many different agencies are engaged in the mission of health, even though their labels may not so indicate. For example, the motor-vehicles department of a state may have a good deal to say about the steps that should be taken to identify an individual on the high-way who is intoxicated or is overcome by a stroke. The scientist wants to say, therefore, that the motor-vehicles department is actively engaged in the health mission. In the same fashion the educational function of the state takes place not only within the department of education, but in many other departments which are engaged in various kinds of training programs for their own personnel and educational programs for the public by means of brochures, short courses, TV demonstrations, and the like. The overall valuation of the education mission therefore cannot take place within traditional department lines.

Why is the management scientist so persistent in talking about missions rather than departments? Simply because by analysing missions he can estimate the worth of an activity for the total system, whereas there is no feasible way of estimating the worth of a department's performance. He needs to know whether one activity of a system component is better than another. But if a department's activity belongs to several larger missions, it may not be possible to distinguish its real contribution. This is why the management scientist is so skeptical about managerial accounting, in any of its various forms. The managerial accountant wants to generate 'scores' of departmental performance, or 'cost centers' which can be examined for their utilization of resources. But insufficient thinking goes into the identification of these scores and centers in terms of their real contribution to the total system objective.

But why do we need components at all? The management scientist would like to look at each choice of the whole system in a direct way, without having to subdivide the choice. But this is not feasible. Consequently, the real reason for the separation of the system into components is to provide the analyst with the kind of information he needs in order to tell whether the system is operating properly and what should be done next. As we shall see, the management scientist thinks he has succeeded reasonably well in certain cases in identifying the real components (missions) of a system.

Unfortunately, to date, in most city and state governments there is no adequate systems analysis of the total system in terms of real components; for historical reasons the state and city governments are divided into departments and divisions that often have no relevance to the true components of the system. As a consequence, says the scientist, the management of our large government systems of states and cities becomes more and more difficult each year. Because the decision making that governs different missions is not centralized, the real missions of the state, e.g. in terms of health, education, recreation, sanitation, and so on, cannot be carried out because there is no management of these missions. One of the greatest dangers in component design is the rigidity which has occurred so frequently in the political designs of the cities and states. The assignment of responsibilities becomes fixed by law and impossible to break. What occurs is a kind of hardening of the communication arteries and the disease that sets in is well known to most administrators. Even the most obvious plans for the various missions of the city and state cannot be carried out simply because there is absolutely no way to break up the rigidities of the system that have occurred because of political history.

It goes without saying that our management scientist is anti-political, simply because so much of politics thwarts the rationality of his designs. He goes so far as to say that city, state, and federal governments cannot be regarded as 'systems', because in their design there is no rational plan of the components of the system and their operation. Nevertheless, there are notable exceptions. Some governmental departments, e.g. the Department of Defense and the National Aeronautics and Space Administration, have taken the 'system challenge' quite seriously, as have several state governments. In industry, 'system thinking' has often infiltrated quite deeply, even though the concepts of the 'whole system' – i.e. the whole corporation – are still very difficult to define. The optimistic management scientist looks forward to a 'systems era' in which man at last will be able to understand the systems he has created and lives in.

The ultimate aim of component thinking is to discover those components (missions) whose measures of performance are truly related to the measure of performance of the overall system.

One obvious desideratum is that as the measure of performance of a component increases (all other things being equal), so should the measure of performance of the total system. Otherwise, the component is not truly contributing to the system performance. For example, in industrial practice if the measure of performance of a component is in terms of its output per unit cost, then it would be essential to show that, as this measure increases, the total performance of the system increases. If, however, drastic cost-reduction methods are imposed on the component that result in decreased quality of its service or product, then it may very well happen that someone has instituted a measure of performance for the component that does not imply an increase in system performance. For example, a production department may institute various kinds of cost-reduction policies resulting in decrease in inventories. Its output per unit of cost may therefore go up, but the performance of the entire firm may go down simply because the cut in inventory leads to unsatisfactory shortages.

These considerations bring us to the last aspect of the system, its management. The management of a system has to deal with the generation of the plans for the system, i.e. consideration of all of the things we have discussed, the overall goals, the environment, the utilization of resources, and the components. The management sets the component goals, allocates the resources, and controls the system performance.

This description of management, however, creates something of a paradox for the management scientist. After all, it is he who has been scheming and plotting with his models and analyses to determine the goals, environment, resources, and components. Is he, therefore, the manager; does he intend to 'take over' with his computer army?

The truth of the matter is that he doesn't want to. He is not a man of action, but a man of ideas. A man of action takes risks, and if he fails, not only does he get fired but his organization may be ruined; the man of action is willing to risk fortunes besides his own. The management scientist is typically a single risk-taker: if he fails, he doesn't have to bear the responsibility of the whole organization's failure.

Hence, we've found one chink in the scientist's armor: he doesn't really understand how he himself is a component of the system he observes. He likes to think that he can stand apart, like the elephant observer, and merely recommend, but not act. How naïve this must appear to the politician is hard to say, but certainly the politician's appreciation of the situation is the more sophisticated one. 'Mere' recommendation is a fantasy; in the management scientist's own terminology, it is doubtful whether the study of a system is a separable mission.

For the moment, we'll forget this embarrassment of the management scientist and talk instead of other ways in which he can aid the managers of systems. Not only does the management of a system generate the plans of the system, but it also must make sure that the plans are being carried out in accordance with its original ideas. If they are not, management must determine why they are not. This activity is often called 'control', although modern managers hasten to add that the term 'control' does not imply strong coercion on the part of management. Indeed, many control procedures operate by exception, so that the management does not interfere with the operations of a component except when the component gives evidence of too great a deviation from plan. However, control does not only mean the examination of whether plans are being carried out correctly; it also implies an evaluation of the plans and consequently a change of plans. As we shall see, one of the critical aspects of the management of systems is the planning for change of plans, because no one can claim to have set down the correct overall objectives, or a correct definition of the environment, or a fully precise definition of resources, or the ultimate definition of the components. Therefore, the management part of the system must receive information that tells it when its concept of the system is erroneous and must include steps that will provide for a change.

The control function of management can be studied by the scientist. The late Norbert Wiener compared this function of the management of the system to the steersman of a ship. The captain of the ship has the responsibility of making sure that the ship goes to its destination within the prescribed time limit of its schedule. This is one version of the overall objective of the ship. The 'environment' of the ship is the set of external conditions the ship must face: the weather, the direction the wind blows, the pattern of the waves etc. From the captain's point of view, the environment also includes the performance characteristics of machinery and men, since these are 'givens' on any voyage. The ship's resources are its men and machinery, as these can be deployed in various ways. The components of the ship are the engine-room mission, the maintenance mission, the galley mission, and so on. The captain of the ship as the manager generates the plans for the ship's operations and makes sure of the implementation of his plans. He institutes various kinds of information systems throughout the ship that inform him where a deviation from plan has occurred, and his task is to determine why the deviation has occurred, to evaluate the performance of the ship, and then finally, if necessary, to change his plan if the information indicates the advisability of doing so. This may be called the 'cybernetic loop'[1] of the management function, because it is what the steersman of a ship is supposed to accomplish. A very critical aspect of a cybernetic loop is the determination of how quickly information should be transmitted. Anyone who has tried to steer a rowboat through rough waters will recognize that, if one responds too quickly – or else too slowly – to the pattern of the waves, he is in real trouble. What is required is an information-feedback loop that permits one to react to the pattern of wind and waves in an optimal fashion.

Wiener and his followers developed a theory of cybernetics which has mainly been applied to the design of machinery. But it is only natural for the management scientist to attempt to apply the theory to the management control of large organizations.

Thus far we have stated the preliminary case for the management scientist's approach to systems, with some critical comments from the sidelines. Does the management scientist's approach work? If 'work' means 'use', then it does indeed work. Hundreds of large industrial firms in transportation, power, communication, and materials all use management science under such labels as 'operations research', 'system science,' or 'system engineering', 'systems analysis' etc. In all cases the avowed purpose of these groups is to approach problems in the spirit outlined in this chapter. Similarly, every section of the military establishment uses management scientists in the design of weapon systems, of information systems (e.g. SAGE and SACCS), of logistics systems, etc. Management science is used extensively in the nonmilitary divisions of the federal government, in public health, in education, in the post office, patent office, National Bureau of Standards etc. Several states and a number of cities are developing management-science capability as an integral part of their government administration.

[1] From the Greek word for 'steersman'.

17 M. D. Mesarović, J. L. Sanders, C. F. Sprague

An Axiomatic Approach to Organizations from a General Systems Viewpoint

W. W. Cooper *et al.*, *New Perspectives in Organization Research*, J. Wiley, 1968, pp. 493–512

Introduction[1]

The principal objective of this chapter is to propose a formal conceptual framework for studies of organizations as systems and to review the results obtained so far which describe the behavior of such systems and shed light on the behavior of organizations.

There are three principal justifications for this attempt:

1. *A formal conceptual model of an organization will allow a more extensive use of formal concepts and analytical or quantitative techniques in studying the behavior of an organization.* This, in turn, will lead to an explanation of many phenomena present in a real-life organization which are as yet unexplainable without a proper framework. We shall touch on some of them in later sections.

2. Organizations are purposeful systems, the elements of which are essentially involved in a problem-solving activity. Now, by the very nature of the decision-making process a model or an image of the object of decision has to exist before a decision is made. (The cases where deliberate simplification of the image is introduced to facilitate the decision-making procedure or make it more economical are also included here.) *A person in an organization has to base his decisions on an image of the rest of the organization, its influence upon him, as well as his influence on the organization as a whole.* In the absence of a proper model, his recourse must be to very crude, often inadequate, descriptions.

It is important to emphasize that a formal model of an organization should help an inside decision maker to properly take into account the structure and functional pattern of the organization. It will not necessarily remove the basic and essential uncertainties

1. This presents the results of research conducted by the Systems Theory Group of the Systems Research Center, Case Institute of Technology, which is supported in part by the ONR Contract No. 1141 (12).

associated with the current and future behavior of the environment and the rest of the organization. The model will allow the decision maker to make the decision with a higher likelihood of success but it will not readily allow 'mechanization' of that process. In other words, 'intuition and experience' of the decision maker will still be needed but the 'logical power' of decision makers will be 'amplified'.

3. *Concepts and methods developed for systems with a given structure cannot be adequately applied to the study of differently structured systems.* The particular case in point is the use of so-called control and decision-making theory for providing an overall model of an organization. For example, occasional attempts have been made to extend some of the concepts of feedback control systems to the study of organizations. These attempts, however, cannot be successful since they are developed for single-level–single-goal systems ($1l1g$ systems, as defined in the next section). They are, therefore, not directly applicable. The lack of an appropriate description of the structure of an organization is a prime reason hindering direct application of analytical and computational methods and techniques to the problems of an organization.

For this reason, among others, an organization is defined in this chapter as a multilevel-multigoal system. In this representation the activities and functioning of the elements of an organization are described in an 'objective' manner as they are seen or felt by the elements, the rest of the system, or the environment. Such an approach does not exclude the role that human nature plays in an organization. It only puts these effects into a proper perspective from which, hopefully, they can be resolved.

Multilevel-multigoal system as a model for an organization

Organization is viewed here from a general systems theory standpoint. Some remarks on the general systems theory seem appropriate.

The general systems theory approaches the study of any system formally, on the basis of how it behaves, without inquiring why it behaves that way or from what fabrics it is made. It is therefore an abstract science.

General systems theory is concerned with abstract models as isomorphisms of systems in the sciences as well as with the classification of these systems with respect to structures, functionings etc. It is only natural, therefore, that the human or social organizations find their place in such a development.

In general systems theory an organization is defined as a goal-seeking system which has interacting goal-seeking subsystems with different goals arranged in a hierarchy. The existence of a multiplicity of goals as a necessary characteristic for a system to be recognized as an organization is widely known and accepted. General systems theory also holds that the existence of a multiplicity of goal-seeking subsystems hierarchically arranged is a sufficient characteristic for a system to represent an organization; that is, the existence of a multilevel-multigoal structure is the primary characteristic of an organization although its actual behavior is ultimately conditioned by the nature and behavior of the goal-seeking subsystem. For example, in the context of human organizations a goal-seeking subsystem in an *mlng* (*m*-levels-*n*-goals) system might represent either a single human decision maker, a group of decision makers, a committee etc.

After one accepts an *mlng* system as a formal model of organization, the development of a new theory for the behavior of such systems as a prerequisite for understanding the behavior of human organizations becomes even more important. This can be illustrated by considering the classification of the goal-seeking systems in the general systems theory. This classification also indicates the relation of different fields concerned with the behavior of formal models of the goal-seeking systems.

Principal classification is in the following three categories:

Single-level–single-goal, 1*l*1*g* systems

A 1*l*1*g* system Figure 1 consists of a causal subsystem S and a goal-seeking subsystem G. The behavior of a 1*l*1*g* system can be completely described in terms of the pursuit of a goal which is either externally or internally generated. The basic features of 1*l*1*g* systems behavior are:

(*a*) Behavior of the *causal system* can be described completely in terms of its terminal behaviour by means of an input-output transformation

$$y(t) = \Psi^* \, x(\tau) \tag{1}$$

where $y(t)$ = output of the causal system

$x(t)$ = input of the causal system

Ψ = systems operator

(*b*) Behavior of the *goal-seeking unit* can be described in terms

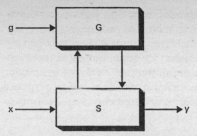

Figure 1 Single-level-single/goal system
S = system unit (causal part);
G = goal-seeking unit; x = inputs;
y = outputs; g = goals?

of its activities in pursuing a given goal. In many cases this can be specified by means of an optimization problem: find m(t) so as to maximize

$$f^* = F\{y(\tau), x(\tau)\} \tag{2}$$

subject to equation 1, where m(τ) represents the action of the goal-seeking unit on the causal system. In situations where uncertainties exist regarding the future behavior of the system or the environment, a 'satisficing' procedure might replace the optimization problem stated above.

Equation 2 specifies the decision-making activity of the system. It is, of course, a very elaborate and complex activity. The functional F which is associated with the system's goal refers to the behavior of the system over a future period of time. This requires, in general, two additional activities which serve to provide the necessary conditions for decision making:

1. Prediction or generalization, in which subsystem G attempts to infer behavior of the system and environment in a future time interval.
2. Learning and adaption, in which, on the basis of the past experience, the subsystem G changes its behavior so as to improve the goal-seeking activity of the system in the future.

$1l1g$ systems offer a conceptual framework for many fields such as control theory, learning and adaptive systems, some self-organizing systems, sequential multistage decision making etc.

Single level-multigoal, $1lng$ systems

A set of interacting $1l1g$ systems represent a single-level–multigoal system (Figure 2). Differences in the goals generate conflict and may cause the development of a competitive situation inside the system. The von Neumann-Morgenstern game theory deals with $1lng$ systems. The theory of behavior of $1lng$ systems is not in a wholly satisfactory state, but game theory at least indicates some of the more important modes of operation. The problems of complex systems in which every $1l1g$ system exhibits adaptation, learning, and prediction needs further conceptual study.

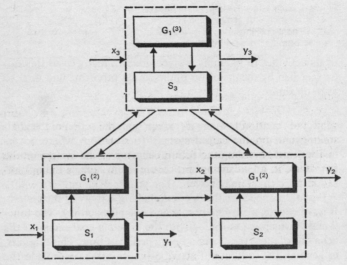

Figure 2 Single-level-three/goal system

Multilevel-multigoal systems, mlng systems

Systems of a still higher complexity are obtained by adding new goal-seeking subsystems which have goals encompassing different portions of the system and may influence the other goal-seeking units. For example, two $1l1g$ systems can be associated with an additional goal-seeking unit which is concerned with a goal common to both of the subsystems and, in fact, to the over-all system (see Figure 3). In order to distinguish the new goal and the activities associated with it, this is termed a 'second-level' goal. This system is termed a two-level–three-goal ($2l3g$) system.

A two-level system may include any number of single-level systems so that, in general, one might have a $2lng$ system. Two second-level systems having one additional common goal with

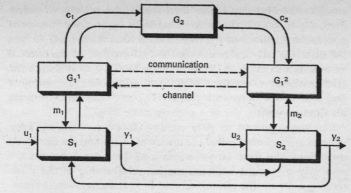

Figure 3 Multiple goal-seeking units

an associated goal-seeking unit represent a 3*l*5*g* system or, in general, a 3*lng* system if any of the two-level systems have more than two subsystems. In the same way one can now build a multilevel-multigoal *mlng* system (Figure 4).

There are advantages to defining more precisely what is meant by 'level' in an *mlng* system. Consider two goal-seeking units G_{ik} and G_{Jl} in an *mlng* system. The goal-seeking unit G_{ik} will be

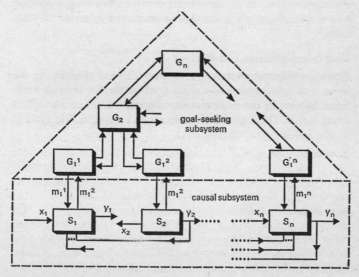

Figure 4 Multilevel/multigoal system

considered to be on a higher level than (have priority of action over) G_{jl} $(i > j)$, if the decision made by G_{ik} directly affects the goal-seeking activity of G_{jl} while the decision made by G_{jl} might influence G_{ik} only indirectly via the performance of the over-all system. For example, a decision made by G_{ik} may influence the goal or objective functional of G_{jl} or it might change constraints and available resources allocated to G_{jl}, and it might even cause an alteration of the decision already made by G_{jl}.

The developments to date coming closest to studying *mlng* systems are the areas called the theory of teams and decentralized operations of a firm. However, both of these developments are primarily concerned with the mutual relationship of the goal-seeking units on the same level and in that sense address themselves to 1*lng* systems. Of principal interest in the theory of *mlng* system, however, are relationships between the units on different levels; i.e. 'vertical relationships' in an organization.

We are making a concentrated effort to develop foundations for a theory of *mlng* systems. In this chapter we shall briefly outline some of the results achieved so far in this direction.

Structure of mlng system

Structure of *mlng* systems can be characterized primarily by the way the goals of different units depend upon each other and how the decisions of a given unit influence the behavior of other units.

Goal interdependence (interaction)

Goals on different levels are, in general, interdependent in the sense that achievement of the ith goal on the jth level depends upon success in the achievement of goals on higher as well as lower levels. These goal interactions can be specified in general by the following set of relations:

$$g_n = g_n(g_1, \cdots, g_{n-1})$$

.

.

.

$$g_i = g_i(g_1, \cdots, g_n) \tag{3}$$

.

.

.

$$g_1 = g_1(g_2, \cdots, g_n)$$

where $g_i = \{g_{ij}\}$ represents a set of goals on the ith level.

To facilitate the decision making associated with different units in a complex *mlng* system, the goal interaction of a goal-seeking unit is almost invariably restricted to the goals in neighboring levels. In the simplest case, the goal of each such unit depends only upon the goals of the first level above it and the first level below it. That is, in the simplest case, equation (3) becomes[2]

$$g_i = g_i(g_{i-1}, g_{i+1}) \qquad (4)$$

Goal interaction is of primary concern when constructing an *mlng* system. Goal-seeking units should interact in a fashion whereby the over-all goal (g_n) is achieved with a reasonable degree of satisfaction. This might require us to specialize the functional g_i so that it monotonically increases with g_{i+1}. This would ensure a motivation for lower level units to further higher level goals.

Two basic types of goal interdependence should be recognized:

Direct goal interdependence. Goals of the higher as well as the lower level units depend explicitly upon the outputs and behavior of the causal systems. The goals of higher level units, however, encompass larger portions of the causal system and, in dynamic situations, may extend over a longer period of time.

Consider for example a 2l3g system (Figure 2). Units G_{11} and G_{12} have goals defined in terms of respective subsets of outputs,

$$G_{11} = G_{11}(y_1, \cdots, y_j) \quad G_{12} = G_{12}(y_{i+1}, \cdots, y_n)$$

The goal of the second level unit, however, depends upon all outputs of the causal system

$$G_2 = G_2(y_1(t), \cdots, y_n(t))$$

Indirect goal interdependence. Goals of a given higher level unit may explicitly depend only upon the goals of the lower level units and implicitly upon the performance of the basic causal system. The relationships specifying the goal interdependence are of the orm given in equations (3) and (4).

Intervention

We now consider how a higher level unit uses its authority to influence the behavior of lower level units. Here we recognize two modes of operation:

Direct intervention. A higher level unit directly changes the

2. Notational convention: the first subscript indicates the level of the goal-seeking unit, and the second distinguishes the unit; thus, for the top level unit the second subscript is omitted.

decision made by the lower level unit unless the latter is found to be appropriate. 'Management by exception' falls in this category. The principal feature of this mode of operation is that the controllability of the causal subsystem that the higher level units possess is often the same as that of the lower level units. However, when intervention is to be realized by changing the decisions at a lower level, the decision-making procedure may become quite complex.

Indirect intervention. A higher level unit cannot directly modify the decision already made by a lower level unit but can modify the goals and constraints of lower level units. For example, in a $2l3g$ system, G_2 may change the rules by which profits will be shared between G_{11} and G_{12} after an operation has been completed. Alternately, it might change the allocation of resources between them. Finally, it might inform G_{11} about the action of G_{12} in such a way that the former's image of the system's operation is changed or even distorted. Here a higher level unit does not have the same amount of controllability as in direct intervention, but the decision-making procedure involved in selecting the best method of modifying the behavior of the lower level unit might be considerably simpler. This mode of operation is found very often in human organizations.

From the brief discussion in this section it should become apparent that the behavior of a goal-seeking unit in an $mlng$ system should be different than the behavior of a unit in either $2l3g$ or $1lng$ systems.

Relation of the general systems organization to social organizations

It is apparent that, in general systems theory, an organization is defined on the basis of the system's functioning irrespective of the presence of the human elements. Two very pertinent questions immediately appear:

1. In what sense is a general systems organization isomorphic to a social organization?

2. How can this isomorphism (if it exists) be used in improving their performance?

The key to answering these questions is, of course, contained in the role which humans play in an organization. Of particular importance are those aspects of its behavior which characterize goal-seeking units. For our purpose we shall classify behavior of a human in an organization in two categories: (1) deductive reasoning; (2) inductive reasoning and chance behavior.

When confronted with a decision-making situation, a human proceeds in the following ways: (1) on the basis of inductive reasoning he builds the assumptions necessary for logical decision making; (2) on the basis of deductive reasoning he makes the choice which appears to be the proper one for the goal as he sees it.

The second aspect of human behavior can be definitely built into a general systems model. How much of the first kind of behavior can be built in is an open question. Some of the effects from the first kind of behavior can be described by means of random processes. Other effects, although present, are irrelevant to the functioning of the organization, or might be made to become irrelevant. Of course, much depends upon the delineation of the two kinds of reasoning or behavior. The limit that divides these two behavior patterns is, in fact, an operational one which moves steadily at the expense of inductive reasoning. Many aspects of human behavior, however, appear to be definitely beyond the limits of deductive contemplation. Such aspects, therefore, seem to remain outside a general systems approach. By the same token, however, those aspects will remain outside of the behavioral sciences and, in fact, any science whatsoever. Therefore, whatever is amenable to scientific method in the field of social studies can be put into a general systems model of an organization.

There is one important aspect of the multilevel systems structure which might help in this situation. Goal evaluation is, in most cases, expressed by functionals which represent many-to-one transformations. There might exist therefore an infinity of action of the lower level units which, if they belong to a given class, will have the same influence on the goal-seeking. A proper use of such nonunique relationships might contribute to making the unknown activities irrelevant.

The second question, that involving the use of an isomorphism, has a more definite answer. Any decision maker involved in the functioning of an organization makes his decisions by using some image of the organization and its relation to the environment. In most cases it is a very much simplified model, and the logical rules which a human uses in decision making are correspondingly simple. A more detailed, even if incomplete, model can only help the decision maker if he is able to use it properly and focus his 'intuition and experience' on the assumptions for this model rather than on the process of building the model itself. The examples discussed in the previous section provide evidence to this end. Built-in redundancy in an organization or in a formulation of the problem again offers great help. In the example studied, the communication channel should not be established for the linear first-level goal no matter what limits are imposed on the actions of first-level-goal units, and no matter how these units evaluate their effort in production processes. The optimal decision is, therefore, made with respect to the class of situations rather than with respect to any particular situation.

Finally, a general systems model should be used which is dynamic and adaptive in the sense that it is brought up to date as knowledge or experience increases. General systems models should, in any event, extend human ability to reason about organizations, just as, by analogy, digital computers extended the ability to calculate. Digital computers can be of significant help here also, but they should be used as logical systems rather than as extended 'slide rules' or 'calculating machines'. The analysis and study of an organization may then be performed on a computer – possibly one of special design – which can perform the optimizations that might be needed when determining the *structure* of an organization.

18 William H. Mitchel

Relevant Neoscientific Management Notions

Excerpt from W. H. Mitchel, *An Approach to the Use of Digital Computers in Municipal Government*, 1968, pp. 136–181

Introduction

In the broad sweep of time, the consideration of factors which contribute to the relative effectiveness and efficiency of institutions has gone full circle. The theory of scientific management originated with Frederick Taylor, whose concern was not so much for the top management view as for scientific management as an attitude and a philosophy. His concepts, when applied, were directed to the operational and first supervisory level. (See Taylor, 1911). In addition, Taylor accounts for some of the notions which were later incorporated into the conceptual framework of a theory of administrative management.[1]

These views were somewhat lacking in concern for the human factors of institutional life. Their underlying assumptions, however, met the realities of the times: jobs were scarce and survival was the primary motivation of the worker. This early body of management theory embraced assumptions about human motivation which were, therefore, primarily economic, though not explicitly so. Human factors, as also determinative of institutional life, were suggested in the classic work of Mary Parker Follet (1941) and confirmed in the empirical research of Roethlisberger and Dickson (1939). Recognition of the existence and significance of human factors in organization, and subsequent research into their nature, could not, in any sense, be said to have joined or supplemented the mechanistic perspective of organizations. Rather, further refinement and exploration of institutional management as a mechanistic process continued, parallel with the new perspectives of psychology, sociology, and anthropology, and the application of their principles to the nature of organizations. Together, these two focal points (or, perhaps, poles of the

1. Fayol, in 1918, first brought into focus an analysis of the managerial function by his articulation of the five elements, and the fourteen principles, of administration (see Fayol, 1949).

spectrum) did not, at any time in the past, form a comprehensive body of theory about institutions, nor do they now (see Massie, 1965). Change and expansion of insights continue to occur in the theory and practice which each perspective provides. The subject of integrated data processing with computers, however, is identified with the perspective of mechanistic theory, rather than with the perspective of organic, socially oriented theory. This is not to deny the value of behavioralism, nor to stand convicted of unawareness of the human factors of organization. Rather, for purposes of exploring one aspect of organization (information) and in exploring that one aspect from only one perspective – that of its rationalization and computer performance – human factors can be isolated and ignored in exchange for the analytic advantages which this limited perspective provides. This approach also requires some mention of the mainstreams of theory which find their genesis in the work of Taylor and his immediate successors. Three additional, major ideas have been added to those ideas of these early scientific management theoreticians, which are of singular importance to both the subject of the research reported by this dissertation and to the field of management theory generally.

The first of these contributions was by Norbert Wiener (1948), with his conceptual notion of feedback and control. The second was that of systems, typified by the writings of the economist, Boulding (1956a), and the biologist, L. von Bertalanffy (1951). Each of these writers was dealing with the dimensions of time and component interaction – the dynamic qualities of institutional life. The third has been a composite of mathematics (including statistics) and the economic notion of marginal utility.[2] This third contribution focused on the ability to quantify variables which affected either institutional goals or the internal factors of organization. Cost-benefit analysis, for example, deals with the goals of the organization and the relative efficiency of various mixes of resources (see Fisher, 1965). PERT, on the other hand, deals with the internal mechanics of institutional operations (Malcolm and Rowe, 1960). Both methods deal with massive problems of

2. This third area is not yet well unified and much of the literature tends to blur necessary distinctions for the temporal advantages of enthusiasm (see Hitch and McKean, 1965). This work is significant in that it is concerned with how to look at goals (defense, in this instance) and how to go about meeting them as a *process* rather than applying substantive solutions to particular problems.

institutional goals and methods of their achievement. Both require quantification and the use of sophisticated mathematical techniques, algebraic and probabilistic, in their applications.

These three major notions constitute an ill-defined return to the philosophy of mechanical precision to which Taylor aspired but at a level of sophistication several magnitudes greater and oriented to a higher organizational level.[3] Most techniques which surround or exemplify these notions are also computer related, or computer driven, because of the extensive use of data and their manipulation. In another sense, each of these new conceptual areas represents the exploration of perspectives for increasing the rationality of man's efforts to engage in collective, cooperative, effort for the achievement of his desires. Rationality, in this sense, is a broad reference to man's efficient use of resources in an effort to achieve his goals.[4] Collectively, these new considerations may be appropriately titled 'Neoscientific Management' notions. They are, primarily, varied attempts to place into theoretical and pragmatic form the scientific management philosophy of Taylor, but with the benefit of a half century of research and experience about what is involved in making the ever more complex organization more efficient.

Viewed from a somewhat different perspective, the ideas of the earlier students of management were based on an atomistic conception of the enterprise rather than an organismic view. The paradigm for the earlier theories was that of the owner-managed enterprise, dealing with his phenomenon entirely from selfish motives, and without reference to the world of which it was a part. This view, whatever its attractiveness then, was probably never true, and certainly the reality of more recent periods has made less applicable whatever truth it did contain. The modern organization, in any event, is more nearly thought of today as resting on a

3. See F. de P. Hanika (1965). An excellent summary of this more recent contribution of ideas for administrative management.
4. 'An action is rational to the extent that it is "correctly" designed to maximize goal achievement, given the goal in question and the real world as it exists. Given more than one goal (the usual human situation), an action is rational to the extent that it is correctly designed to maximize *net* goal achievement. When several actions are required to attain goals, rationality requires *coordination*; that is, the actions must be scheduled and dovetailed so that net goal achievement is not diminished by avoidable conflicts among the actions' (Dahl and Lindblom, 1953, p. 38). This work is a remarkably fine discussion of the basic social processes which attempts at rationality generate.

whole series of networks and relationships, both within the institution and external to it.[5]

Because simplistic views of the organization no longer dominate the underlying conceptual structure which influences man in the creation and management of his institutions; because of the more recent needs and attempts to systematize the complexity of institutional life which he now perceives; and because various conclusions about how to achieve a greater level of rationality in institutional performance are operative upon all institutions, including the municipalities, the structure of these neoscientific concepts is relevant to this discussion. Indeed, the notion that a single integrated management information system could be developed is, itself, both a product and a reflection of these newer ideas. The conceptual roots of the research which fostered this dissertation are, themselves, to be found in such neoscientific notions as systems, quantification, feedback and electronic data processing. [This] research project was an attempt to develop an empirical and concrete manifestation of these neoscientific theories about the management of institutions. Further, if these manifestations are to be meaningful and have utility, they, in turn, must be consistent with the thrust of these newer rationalistic notions, for their impact will be felt as much by the municipalities as by the other institutions which populate an industrial society and which compete with it for the limited resources.

There exists no satisfactory typology for these recent neoscientific developments. Indeed, it is not unusual to find that under one scheme of classification a given theory (systems, for example) will be considered the set, while in another, the same theory will be considered a subset. Further, the terms used to designate the theories vary as to what is, or is not, subsumed under the category title.[6] No claim is made for an exhaustive listing of the newer notions, nor for full exploitation of the ramifications which each notion represents. What appears necessary to the present discussion is to bring into focus five of the neo-

5. For an interesting discussion of these notions, see Maslow (1965). See also Bennis (1964).

6. The difficulty of reducing to any meaningful form the complex and constantly changing body of knowledge about organization is enormous and beyond the scope of this paper – and probably beyond the talents of the author. Clarity and length are forever at war. The problem of clarity is also confounded by the lack of structure in the management sciences. See, for example, Koontz (1961).

scientific notions; these are selected because of their immediate relationship to the research goals and ideologic reference structure.[7]

The systems approach

Perhaps the most basic and far reaching of the neoscientific management concepts has been the application of general systems theory to the organization. The initial formulation of the concept of systems is undoubtedly lost to antiquity and confounded by the many forms its expression can take. In the more recent past, the expression of the underlying notions of general systems theory are found in the works of at least three men: von Bertalanffy, Cannon, and Einstein.[8] The most fundamental and philosophical justification for the systems approach, as well as its intellectual appeal, stems from the basic observations of Einstein, who suggested that there was no phenomenon of instantaneousness. Rather, the universe is a complex of non-simultaneous and nonidentical events. It follows from this view that there can be no possible unit and static conceptual picture of the universe. Rather, the universe and its components are in a continuing process of transformation in a nonsimultaneous sequence, as a function of time.[9] Stability, therefore, is to be found, if at all, in the processes of change and interaction, rather than in structures of simultaneously existing phenomena.

From such abstract conjectures about the problems of phenomena in time streams, permanence, and the process of change there has developed a body of thought which in its organized form is called 'general systems theory' (Boulding, 1956b). The central goal of such theory is to develop some orderly way of handling the constantly shifting, interacting individual events of existence. Definitions of general systems theory, as well as its correlative propositions, leave much to be desired in precision or applications potential – as attractive as the notions

7. A more detailed consideration of these developments under a different typology is contained in Hanika, *op. cit.*

8. A biologist, a medical researcher, and a physicist, all of whom observed and struggled with the problem of structure and change in time. Ludwig von Bertalanffy (1952); Cannon (1963); and Einstein, as reported by Buckminster Fuller (1966).

9. There is an interesting parallel between Newtonian physics and administrative structural theory, and Einstein physics and systems theory of organization. One deals with an interlinked mechanical universe; the other deals with an interacting universe in organic change.

appear at first blush. For example, one accepted expression of the theory is as follows:

Thus, there exist models, principles, and laws that apply to generalized systems or their subclasses, irrespective of their particular kind, the nature of their component elements, and the relations of 'forces' between them. It seems legitimate to ask for a theory, not of systems of a more or less special kind, but of universal principles applying to systems in general. In this way we come to postulate a new discipline, called General System Theory. Its subject matter is the formulation and derivation of those principles which are valid for 'systems' in general (von Bertalanffy, 1952).

Other definitions contribute specificity, but at the price of extending even further the scope of the theory, as Boulding does when he posits a hierarchy of complexity as part of general systems. Agreement about the referents of the theory – if, indeed, it is a theory–appears lacking.[10] Further, because intellectual tools which attempt to provide an ordering framework for phenomena in process of continuing change are both needed and useful, the theory has been modified and applied in a variety of contexts with or without regard for the precision of formulation which adequate theory requires.[11]

The same lack of precision which characterizes basic systems theory also is found in specific applied research areas.[12] In most instances, the applications reflect a reaction against the growing compartmentalization occasioned by specialization and its complexity. These varying uses of systems theory also reflect the need to reintegrate analytic segments into operational units.[13] Essentially, the process of operationalizing general systems theory to a specific referent field appears to involve selective differentiation of both the theory and its correlates.[14] Use of the process in the field of organization and management characterizes this

10. Much of the discussion about systems which appears here has been taken from the following four major works which sketch out and deal with the notions of the systems approach as a concept, as theory, as a model, and as an analytical approach: Blalock and Blalock (1959); Kuhn (1965); Chin (1962); Balding, loc. cit.

11. McClelland (1962) denies that it is a theory and, instead, suggests that it is a certain point of view!

12. Richard A. Johnson et al. (1967).

13. Human ecology movement in environmental research is an excellent example.

14. For an excellent example of systems theory used in the development of technological artifacts (communications), see Harold Chestnut (1966).

selective process.[15] In the specific area of organizational behavior, perception of the organization as a single system has great utility. This utility is found whether viewing the organization internally, or in relationship to its environment. Further, the notions of systems can be arranged to serve different purposes depending upon the interests or goals involved. For example, an institution may be viewed as a concrete model (i.e. an existing operating system) to determine, empirically, what the interacting elements are. Or it may be viewed as an analytic model, i.e. a constructed simplification of some part of reality which retains only those features regarded as essential for establishing the pattern or assumed important relationships.[16]

Gross, in viewing systems theory in relationship to organizations, suggests that systems may be divided into 'structural' and 'performance' systems. The structural system, according to this view, consists of:

1. People . . .
2. Nonhuman resources.
3. Grouped together into subsystems.
4. That interrelate among themselves.
5. With the external environment.
6. And are subject to certain values.
7. A central guidance system that may help provide the capacity for future performance.

The performance elements include:

. . . activities to satisfy (1) the interests of various 'interesteds' by (2) producing various kinds, qualities, quantities of output, (3) investing in the system's capacity for future output, (4) using inputs efficiently, (5) acquiring inputs, and doing all the above in a manner that conforms with (6) various codes of behavior and (7) varying conceptions of technical and administrative (or guidance) rationality. (Gross, 1966.)

Within these social systems, which are concepts for viewing institutional life, lie even more limited notions about the components of organizations which exist, or are consciously created by man, for achieving specific goals (not the least of which is the creation of further goals!). This view sees the organization as a

15. See, for example, David S. Stoller and Richard L. van Horn 1958. Contrast the Stoller–van Horn approach with E. R. Dickey (1963). Dickey defines systems in terms of application to an automated information system. See also John A. Seiler (1967) and Miller (1965).
16. See R. Chin (1962).

dynamic entity with activities which interact, require coordination and control, are concerned with survival and change, and which form one element of the economic, technical and social system of the nation. The systems approach at this level of analysis tends to incorporate notions of boundaries and interaction, and other concepts which are equally abstract and difficult to reduce to concrete terms: control, stability, growth, rationality, and choice, to mention a few. It is not clear whether these additional notions are essential to the viability of the underlying systems concepts, or whether they are required analytic devices to articulate the systems aspects of organization. What does appear to be clear is that there is in process of formation a congeries of notions about the organization as a *system* which can be subjected to analysis, empirical study, description and conscious manipulation.[17]

Some current major notions about the institution as an operating system are given. First, the organization as a system contains a hierarchy of subsystems. These subsystems have boundaries just as the institution, as a system, has boundaries with the community of which it, in turn, is a subsystem. Within the boundaries, there are interdependent forces at play in intimate relationship, producing a cumulative total effect on the organization. This combined effect creates a continuing change in the relationships between the system and other systems, which share common boundaries. The advantages of using the subsystem approach include the ability to avoid the overwhelming complexity which an organization represents. It permits a relatively limited number of events to be identified as taking place within a particular system and a limited number of parts of the system.

Second, there is an interaction between the social forces within an organization, on a continuing basis, and within some framework which displays a greater permanence than the specific interactive events. Within this interactive pattern, the interests of the institution and its members are not thought of as being identical, nor are the interests of the members of the organization identical, nor necessarily similar. Rather, many interests are simultaneously in the process of being served and articulated.[18]

Third, relations between the subsystems, the individual members, and the organization, tend toward a steady state, or a balance of forces, which is relatively enduring. This relative

17. Seiler's work is an example of such a grouping and evaluation of their use.

18. For an interesting discussion of this point, see R. Likert (1961).

equilibrium tends to be dynamic or moving, which causes constant changes among the relationships and consequent new levels of stability. Also included in this notion is the phenomenon of two tendencies – that of striving toward an established stability within the organization and that of improving the stability between the organization and its environment.

Fourth, the system may be conceived of as having entries through its boundaries (input), modification of those entries by the internal capacity of the system (processing, storage, etc.), and the emission of products from the organization (output). This simplistic input-output model becomes the basis for the measurement of relative efficiency and rationality of organizational activity.

Fifth, because the output of a system has an impact on the organization's environment, and is the result of the input plus internal processing, and because the organization has apparent strong tendencies to survive, a continuing part of the input and processing relates to indicators of what the external environment's reaction to the output is, and what internal conditions affect the input-output process. This 'feedback' process characterizes all durable organizations. In a sense, the feedback process conflicts with the tendency to equilibrium within the organization.[19] The system configuration and its boundaries are in a constant state of change as this feedback from within and without the system indicates a constantly imperfect adjustment of the system, both within itself and with other systems.[20] The systems approach to organization does not require a perception of harmony, but of conflict in suspense. Both the system and its dynamics are in constant process of change as the result of explicit, rational, action exercised by its goal-seeking members as a result of feedback reflecting their imperfect integration, the system's environment, and individual interests fulfilled, or not fulfilled.

Sixth, continuing, viable organizations appear to require certain key subsystems to insure the informational and decision system which input/output and feedback processes both generate

19. Optner suggests that the structure through which 'input' is converted to 'output' must be continually changed on the basis of 'feedback'.

20. Theoretically, there are two types of systems changes: stationary equilibrium, where there is a fixed point to which the system returns after a disturbance; and dynamic equilibrium, where the equilibrium shifts to a new position of balance after a disturbance. Probably no organization displays more than a general approximation to the stationary equilibrium model.

William H. Mitchel 313

and require. Johnson, Kast, and Rosenberg have suggested the following as essential:

1. A sensor subsystem designed to measure changes within the system and with the environment.
2. An information processing subsystem such as an accounting, or data processing activity.
3. A decision-making activity which receives information inputs and outputs planning messages.
4. A processing subsystem which utilizes information, energy and materials to accomplish certain tasks.
5. A control or feedback subsystem which measures plan to accomplished process.
6. An information storage subsystem which may take the form of records, computer programs etc.[21]

Because some of the major characteristics of an organization as a system have been articulated, it is possible to approach the analysis of an organization from this perspective: the 'systems approach', using processes which have become known as 'systems analysis'. Kraemer has suggested that four key ideas make up the contribution of systems analysis to organizations: the black box or computer, the system environment or boundary characteristics, system stability and system stress or tension, and factorization and synthesis. It is probable that systems analysis, while using these key ideas, has also included a wider spectrum of analytic techniques which, in varying degrees of proximity, relate to the creation or improvement of organization, and are based on the assumption that it is a system. The following sections of this chapter will deal with the more important current theories. The discussion is designed to bring into sharper focus the boundaries of a system, how a system operates, and how an organization can be made to achieve its goals more efficiently, as well as the goals of its constituent members. The systems analysis process, in its current state of development, seems to include the following:

1. A sequence of activities directed toward a given organization or program.

2. Definition and detailed description of the boundaries of the system under consideration.

3. Functional description of the system in terms of component subsystems and their operational interactions.

21. R. A. Johnson, F. E. Kast, and J. E. Rosenberg (1964).

4. Determination of objectives and criteria of optimal system performance.

5. An examination of reasonable alternative configurations of system elements that approximate optimal system performance, and a determination of the consequences of each of these configurations in terms of its feasibility, acceptability, and cost effectiveness.

6. An objective or quantified presentation of these alternatives, and supporting evidence, to the managerial group for their evaluation and selection of an alternative for actual implementation.[22]

The sequence described above is deceptively simple in concept and exceedingly difficult to apply. It requires not only that the group approaching the problem be interdisciplinary in their training, but that they form an integrated team 'acting in a common environment on a day-to-day basis with a common problem-oriented focus.' It would seem that if the organization is to be viewed as a system, then the methodology for its study and analysis must also be shaped in the mould of the systems model.

Systems analysis is an eclectic technique, resting upon the notions that the phenomenon of interest can be viewed in its total, as a system, and that many ways for improvement can be expressed in terms of their relative merit and cost. The critical element in the process is to define what features of the system are of particular interest and what boundaries are to be placed on that system. For purposes of the research work undertaken by the Municipal Information Systems Project, for example, the boundaries constituted the city as an organization and as a community. The system perspective was the flow of information and its relevance to decisional processes. No consideration was given to the behavioral system, the authority system, the political system, or the other system perspectives which constitute the municipality. The project used as its point of reference the hypothesis that the city is an information system. Research efforts could then be directed in examining the phenomenon of 'city' to determine if it met the criteria of a 'system'.

22. The list is a paraphrasing of a statement before a Congressional Subcommittee interested in the systems analysis process, by Thomas C. Rowan, Vice President and Manager of Advanced Systems Division, Systems Development Corporation, Santa Monica, California, U.S. Congress, Senate, Special Subcommittee on the Utilization of Scientific Manpower, Committee on Labor and Public Welfare, *Hearings on S. 2662*, 89th Cong., 1st and 2d Sess., 1966, pp. 161–171.

At least four major problem areas apparently limit and condition the use of rationalism as a process for the institutional manager[23]:

Information: The number of decisions which can be made on the basis of full knowledge of the choices of cost/benefit mixes which are either theoretically or practically available, is comparatively small and usually confined to routine procedures. Where social questions are involved, the level of quantified indicators hardly justifies optimism in any alleged precise definition of the important variables.

Communication: Assuming that some reasonable grasp of the variables could be secured for a given problem facing the organization, the problem of communicating both the quantified factors, and the underlying assumptions, to the decision centers is relatively unexplored.

The number of variables: There is evidence that the individual can hold only a limited number of variables in his mind at any one time. Yet, the very assumption of economic rationalism includes the notion of a simultaneous comparison of a large number of combinations, and their comparison, for relative merit. The introduction of the computer with its ability to handle larger numbers of variables has eased this problem. However, the larger problem of quantifying these variables appears less amenable to easy solution.

The complexity of the relationships among the variables: Not only is there a multiplicity of variables, including those involving value preferences; there is also the fact that these variables are not linear. Rather, they interact to form vectors of an almost infinite variety and direction. Sophisticated mathematical techniques for the orderly handling of such interactive processes are not available.[24]

The limitations on administrative rationality, as well as its distinction from the normative theory of economic rationality, has been central to the growth of theory about rationality under

23. This typology has been borrowed from Dahl and Lindblom, *op. cit.*, where it is used in a different context. It appears applicable, however, to the more limited scope of this discussion.
24. The limits are roughly suggested in the limitations which operate in game theory where only alternatives can be accommodated in uncertainty situations. See M. M. Flood (1962).

conditions of uncertainty.[25] Simon, one of the most significant contributors in the field of administrative rationality, has suggested the principle of 'bounded rationality' as an alternative to the normative conduct of economic man where it is assumed that all knowledge of all possible choices is available. Man, under the Simon model, is assumed to construct a simplified model of the actual world, and then is expected to act reasonably in terms of that model. A further limitation is also made by Simon, in that he accepts as significant the factor of limitations on the energy available to man for 'search' for possible solutions to problems. Simply, man searches only until he finds the answer which appears the best of those readily available and consistent with apparent needs. The limits placed on administrative rationality by inadequate knowledge and the human thinking process results in what Simon calls 'satisficing'.[26] Attempts to bring administrative rationality closer to the rubric of economic rationality apparently continue to require the assumption of an ordering of value preferences, recognition of certainty as against uncertainty, and the assumption of stability during which the selection process occurs.[27]

The element which repeatedly recurs in any selected set of readings on the problems of administrative rationality is that of information and its quantification, either on an ordinal or a digital scale. Rationality, seemingly, is grounded in the notion of preference (explicit merit), and articulated where the preferences can be expressed numerically. Such preferences, to be meaningful, or brought into sharp focus, must rest on some orderly manner of representing the conditions under consideration, and their

25. Simon, since his initial work (1947), has consistently distinguished between economic and administrative man. 'Traditional economic man, however attractive he is to the economic theorist, has little or no place in the theory of organization. . . .' Simon, *Models of Man*, p. 198.

The British cybernetician, Stafford Beer, has divided *all* systems into 'deterministic' and 'probabilistic', which bears a rough equivalent to the distinction made here. Deterministic systems involve complete knowledge of system parts reaction – a computer would be an example of such a system. Probabilistic systems involve relative degrees of predictability based on the availability of facts and analytic techniques. Stafford Beer (1959).

26. 'The key to the simplification of the choice process . . . is the replacement of the goal of *maximizing* [the economic man goal] with the goal of satisficing, of finding a course of action that is "good enough" [the administrative man goal].' *Ibid*.

27. See L. J. Savage (1951) and W. Edwards (1961) pp. 473–98.

relative character or magnitude. Because the phenomenon of urban life can, in varying degrees, be represented in quantified form, and because the computer, as a general purpose symbol manipulator, can readily process quantified information, there is, in the union of computers to administration, the basis of increasing administrative rationality. Rational behavior appears central to both the theory and the practice of administration.

Cybernetic control patterns

The term 'cybernetics' was coined by Norbert Wiener to describe a process for maintaining systems stability in electrical circuits, and later expanded to the more general phenomenon of control and communications in 'animal and machine' (Wiener, 1961). While the work of Wiener is mathematically oriented, the matrices in which the concepts of cybernetics were developed included the fields of medicine and computers.[28] However, perception of the nature of the phenomenon of feedback was discussed and contemplated by at least two other and earlier scientists: Walter B. Cannon and Jacob von Uexkull, both of whom were involved in life science research and had perceived the self-regulating characteristic of organisms.[29] Wiener also observed that 'the social system is an organization like the individual, that it is bound together by a system of communication, and that it has a dynamics in which circular processes of feedback nature play an important part. . . .' It was this latter sensitivity which occupied a good portion of his later writing and exploration and is – in part – responsible for the development of a body of thought

28. The introduction to the first edition contains an interesting history of the personages in both fields who participated in the cross fertilization of ideas and scientific methodology which is part of the heritage in conceptualizing the process of 'feedback.'

29. Cannon sought to generalize the interdependence of the respiratory, circulatory, gastrointestinal, and the glandular systems into a control pattern involving feedback. Von Uexkull termed the phenomenon 'the function-circle' with a perception and an action segment. It would appear that von Uexkull not only considered rote feed-back, but also the possibility of processing prior to output – the selective processing based on stored information about conditions relevant to the input, and in terms of which the accommodation becomes highly selective. It is interesting, also, to note that Dr Cannon was a colleague of Dr Rosenblueth who, in turn, was a colleague of Dr Wiener. Dr Wiener and Dr Cannon apparently had also discussed the generalized expressions of the phenomenon. See also Robert McClintock (1966). Von Uexkull's work is apparently not available in an English translation.

about the feedback process.[30] The significance of Wiener's work falls somewhere between that of an elemental recognition that some mechanisms are self-regulating while other are not and the creation of an analytic theory which not only accommodates the gross facts of feedback but lays the framework for the construction of new systems.

One of the distinguishing characteristics of a continuing system is that it possesses working components which maintain typical processes despite variation, both within the system and in terms of the environment in which it exists. When this adaptive behavior is ineffective, the organism ceases to exist. Essential to this adaption is the use of some of the energy available to the organism to regulate the amount of energy expended, and to focus its direction. The analogy between the municipality and its environment, and the more generalized 'organism' is important. However, to realize an adaptive situation, three conditions must be met:

1. The changes required must be controllable by some physical means, i.e. a regulating organ.

2. The phenomenon, to be controlled, must be measurable or comparable with some external standard.

3. Both the regulatory mechanism and the measurement must be rapid enough for the task.[31]

In addition, any feedback system is characterized by oscillatory behavior due to the 'noise' in the feedback system and inertia, which the characteristic effort to correct discrepancy displays. In the social system area, another factor also insures a continuing imperfect adjustment: the fact that the social system is an adaptive rather than a recursive one. Simply, the measurement for readjustment changes during the process of correction and also reflects the effort to readjust.

The natural social systems, i.e. those that have not been explicitly organized and established by the conscious action of humans, need not have a cybernetic quality about them. To the contrary, because they tend to be open systems (no a priori goal

30. It is difficult to separate some of Wiener's notions from the electronic computer which has made some of his mathematics operational. He was also interested in the extrapolative potential of the cybernetic/computer union, particularly in the area of man-machine interface. See Norbert Wiener (1964). An interesting discussion of the impact of the cybernetic model on modern culture is contained in Harvey Brooks (1965).

31. Arnold Tustin (1955).

definition) and almost unsystematic, they may not have any feedback such as characterizes the individual organism. However, the moment there is a conscious, explicit effort to order some future set of social conditions, there is need for a cybernetic system.[32] In its absence, there appears no possibility for insuring that interim events between the explicit expression of a plan, and its scheduled achievement, will not render the plan unattainable. Further, the plan itself, as well as the mechanics for its achievement, requires continuing evaluation as the environment in which it is gestating changes or seeks a more desirable future status. In specific terms of the municipality, there is increasing evidence to suggest that the relatively open system of urban development is to be subjected to more explicit control through reliance on the then increasing comprehensiveness. For the expanded planning process to be successful, some measure of its achievement, as well as information on the environment in which the plan is to become operative, becomes a continuing requirement; feedback is an essential of planning.[33] Feedback, in this limited sense, refers to the total municipal information process through which primary and second-order effects of the city's organizational actions are fed back to the organization and compared with desired performance.[34]

The application of feedback techniques and theory to the

32. The distinction between 'cause' and 'effect' is largely subjective, and relates to the explicit recognition of man that he not only can visualize some more desirable future state of events, but also can take steps to increase the probability that future events will approximate his desires. 'Cause', in this sense, is what might be conceivably manipulated, and 'effect' is what might conceivably be proposed.

The entire process of 'planning', therefore, relates to considerations of conceivable futures, and the assumption that sufficient knowledge exists by which manipulative events can be arranged to result in and approximate a future desired state of affairs.

33. Essentially, any feedback involves an arrangement whereby the output from a process or effort is measured and compared to a preestablished goal or standard. Where significant deviation occurs, corrective action is instituted. Communication of output (performance) for processing, and modification of input (control) are elemental to the feedback concept. When the process is achieved without manual intervention – 'automation' – one form of automation is achieved. The computer, in this situation, is the handmaiden of automation. The communication of output measurement, and its comparison to a standard, normally involves digitalizing the information, i.e. reducing the data to symbols. The symbiosis of computer and feedback processes is obvious. The conceptual interdependence of systems concepts and feedback is also obvious.

34. Robert A. Rosenthal and Robert S. Weiss (1966).

empirical world of municipal government involves at least seven major problems:[35]

1. Feedback and organizational survival. Governmental institutions survive in terms of their function in the larger environment of which they are a part. Continuing information about the demands and variables of that environment is, therefore, necessary for survival. The difficulties in determining what factors in the environment are relevant to institutional survival are ill-defined.

2. Estimating feedback requirements. Theoretically, all output should be measured and compared against a standard or goal. As a practical matter, only a very limited portion of an organization's output can be captured and reduced to the specific terms required for evaluation. Further, as among the practically available values, there is very limited knowledge as to what values are useful, either as indicators or as standards.

3. Feedback from external sources. The city forms an ill-defined subsystem within the urban system of both public and private institutions and individual citizens. The problem of feedback from these external sources includes knowing what information is available from such sources and its relationship to the operations of the municipality.

4. Feedback from internal sources. In general terms, this form of feedback approximates the traditional 'management information system', and has been primarily concerned with fiscal type data. However, as there is expanded understanding of what variables need to be taken into account to measure the relative efficiency of the institution, fiscal feedback proves correspondingly less satisfactory in revealing the conditions of the institution.[36]

5. Feedback and its effects on policy. The relationship of information about performance or external conditions to the decisional processes of the municipality has not been systematically investigated. There is an unproven assumption that policy centers of the municipality will react in some rational fashion to improved (i.e. more and accurate) data about either the community or the municipal organization. However, the assumption approximates an unproven hypothesis.

6. Dangers of organizational feedback: Simplistic views tend to prevail concerning the effect of availability of information on

35. The typology, but not the discussion, is essentially that of Rosenthal and Weiss (1966).
36. See Warren G. Bennis (1962).

institutional performance. However, the extrapolative power of a single unfortunate incident is well known. Further, feedback data are fitted into the pre-existing images of the recipients where a variety of both anticipated and unanticipated responses can be generated. Feedback data enjoy no greater 'objectivity' than do all other types of information.

7. Research in maintaining feedback. This is primarily a problem in the linear value of whatever feedback data are collected. Standards against which feedback data are compared are, themselves, often the product of aggregation over time, and a normative derivation from that summary. But both the focal points of interest within the organization and the phenomenon being measured continue to change. As the measurement units drift away from their initial referents, or as the phenomenon being measured changes its essential nature, the value of the feedback deteriorates. If continuing research is carried on to insure measurement of currently significant phenomena, then both the units of measurement and the phenomena measured will change. With the change, comparability over extended time periods becomes meaningless.

The concept of cybernetics is a powerful tool in the arsenal of administration. The surprisingly simple structure of the concept, however, is matched with an equally surprising complexity in its application to the problems of modern organization. Where processes are relatively simple and standards readily measured, the control leverage afforded by its application is great. But where the processes are complicated by a mixture of human values, aspirations, different communication media, and cultural subtleties, the application difficulties appear formidable.[37]

37. To say nothing of the emerging perception that 'environments are not just containers, but are processes that change the content [of language] totally. New media are new environments . . .' McLuhan (1966).

References

BEER, S. (1959), *Cybernetics and Management*, London University Press.
BENNIS, W. G. (1962), 'Towards a "truly" scientific management: the concept of organization health', *General Systems yearbook*.
BENNIS, W. G. (1964), 'Organizational developments and the fate of bureaucracy', invited address, Division of Industrial and Business Psychology, AMA.
BERTALANFFY, L. VON (1951), 'General systems theory', *Human Biology*
BERTALANFFY, L. VON (1952), *Problems of Life: an Evaluation of Modern Biological Thought*, Harper and Row.

BLALOCK, H. M., and A. B. (1959) 'Towards a classification of systems analysis in the social sciences', *Phil. Sci.*, XXVI.

BOULDING, K. (1956a) 'General systems theory: the skeleton of science', rewritten for *General Systems*, I.

BOULDING, K. E. (1956a), 'General systems theory: the skeleton of a management science', *Man. Sci.*, April.

BUCKMINSTER FULLER, R. (1966) 'Vision 65: summary lecture', *The American Scholar*, Spring.

BROOKS, H. (1965), 'Scientific concepts and culture change', *Daedalus*, XCIV.

CANNON, W. B. (1963), *The Wisdom of the Human Body*, Norton.

CHESTNUT, H. (1966), *Systems Engineering Tools*, J. Wiley.

CHIN, R. (1962), 'The utility of systems models and development models for practitioners', in W. Bennis *et al.* (eds), *The Planning of Changes*, Holt, Rinehart and Winston.

DAHL, R. A. and LINDBLOM, C. E. (1953), *Politics, Economics and Welfare*, Harper and Row.

EDWARDS, W. (1961) 'Behavioural decision theory', *Ann. Rev. Psych.*

FAYOL, H. (1949), *General and Industrial Management*, Pitman.

FISHER, G. H. (1965), 'The role of cost-utility analysis in program-budgeting', in D. Novick (ed.), *Program Budgeting*, Harvard University Press.

FLOOD, M. M. (1962), 'A symposium on game theory', *Behav. Sci.*, VII, no. 1.

FOLLETT, M. P. (1941), *Dynamic Administration*, Management Publications Trust.

GROSS, B. (1966), 'The state of the nation: social systems accounting', in R. Bauer (ed.), *Social Indicators*, MIT Press.

HANIKA, F. DE P. (1965), *New Thinking in Management*, Hutchinson.

HITCH, C. J., and MCKEAN, R. N. (1965), *The Economics of Defense in the Nuclear Age*, 4th edn, The RAND corporation.

JOHNSON, R. A., KAST, F. E., and ROSENBERG, J. E. (1964), 'Systems theory and management', *Man. Sci.*, January.

JOHNSON, R., *et al.* (1967), *The Theory and Management of Systems*, McGraw-Hill.

KOONTZ, H. (1961), 'The management theory jungle', *Jour. Ac. Man.*, December.

KUHN, A. (1965), *The Study of Society: a Unified Approach*, Irwin.

LIKERT, R. (1961), *New Patterns of Management*, McGraw-Hill.

MALCOLM, D. G., and ROWE, A. J. (eds.) (1960), *Management Control Systems*, J. Wiley.

MASLOW, A. H. (1965), *Eupsychian Management*, Irwin.

MASSIE, J. L. (1965), 'Management theory', in J. March (ed.), *Handbook of Organizations*, Rank, McNally and Co.

MCCLELLAND, C. (1962), 'General systems and the social sciences', *ETC: Rev. Gen. Sem.* XIX.

MCCLINTOCK, R. (1966), 'Machines and vitalists', *Amer. Schol.*, Spring.

MCLUHAN, M. (1966), 'Address at Vision 65', *Amer. Schol.*, Spring.

MILLER, J. (1965), 'Living systems: basic concepts', *Behav. Sci.*. July.

OPTNER, S. L. (1960), *Systems Analysis for Business Management*, Prentice-Hall.

ROSENTHAL, R., and WEISS, R. (1966), 'Problems of organizational feedback processes', in R. Bauer (ed.), *Social Indicators*, MIT Press.

ROETHLISBERGER, F. J., and DICKSON, W. (1939), *Management and the Worker*, Harvard University Press.

SAVAGE, L. J. (1951), 'The theory of statistical decision', *J. Amer. Stat. Ass.* XLVI.

SEILER, J. (1967), *Systems Analysis in Organizational Behaviour*, Irwin.

STOLLER, D., and VANHORN, R. (1958), *Design of a Management System*, The RAND Corporation.

TUSTIN, A. (1955), 'Feedback', *Automatic Control*, Schuster.

UEXKULL, J. VON (1926), *Theoretical Biology*, Harcourt, Brace & Co.

WEINER, N. (1948), *Cybernetics*, J. Wiley.

WEINER, N. (1964), *God and Golem, Inc.*, MIT Press.

19 Simon Ramo

Systems Approach to Man's Future

S. Ramo, 'Systems approach to man's future', *Los Angeles Times*, 1970

Along one river in the United States, several cities and numerous plants depend on the river's water supply and pour their waste into it. Vessels operating busily on the waterway contribute their share of contamination. The river long ago lost its utility for recreation. A portion of its surface actually caught fire recently.

A similar situation exists in another area where a highly industrialized population pollutes an ocean bay.

In the first example, an awakening citizenry decided to attack the problem. They assembled a team of professional experts and asked them to analyse every facet of the water pollution problem. This group included scientists, engineers, businessmen, economists, academic sociologists, and practical politicians.

They began by raising questions. Should the goal be the greatest industrial expansion? What were the inevitable conflicts amongst the separate interests involved? Could the region's economy support antipollution measures? How should the criteria of an 'unpolluted' river be defined?

They knew that no one can say how positively he wants something unless he knows how much he is going to have to pay for it. Nevertheless, in a reasonable time, the citizens were shown detailed predicted consequences of continued pollution: deterioration of health, lower economic growth, poorer living standards. Also pictured was what it might take in commitment of resources, handicaps to industry, and bigger bond issues for new standards to reduce pollution and thus realize a superior environment.

Agreement reached

Everyone understood that the analytical simulation of the alternatives, ranging from inaction to different actions, were only estimates. There were unresolved differences of opinion, people being people, their interests diverse. But some ballpark comparisons could now be pondered, and the fundamental proposal received majority support. So they chose to go for a major program to unpollute that river. The several cities agreed to set up a higher authority empowered to control its use.

Meanwhile, at that other location around the common ocean bay, some called for a similar approach, but more were loudly apprehensive about conceivable results.

Some politicians (and a few professors) shouted that an analytical approach was good only for landing a man on the moon. A company pointed out that, if forced to treat output waste fully, its costs would go up, competitors would take its business, and unemployment would result. One mayor said, 'antipollutionists' were promoters looking for profit. Another claimed it was the other cities that were ruining the waters. And in the end, nothing was done.

Now, both of these examples are fictitious, but each has foundation in reality. The second situation, inaction and social immaturity, is the rule. The first example, going ahead and doing what should be done, is a rarity. However, basic science and methodology are, or can be made, available. Furthermore, the good of the society militates towards the use of a scientific, i.e. a logical, objective approach. The reason for the rarity is man's missing will, not a lack of means.

Accelerating technology is badly mismatched in our time to lagging social advance. Hence, we see about us not only the problems of misused technology.

Granted that our unsolved 'social engineering' problems are much tougher than purely technological projects because of the complex human element, let us ask specifically: Have the big defense and space projects taught us something that we can carry over to pollution, transportation, education, medical care, urban development? In the ICBM and Apollo programs, have methods been developed which can be used to accelerate our handling of these urgent social-engineering problems?

The space program has indeed taught us something about how goals can be decided, but something not necessarily useful, I fear, when applied to down-to-earth problems.

The lunar program was launched as a reaction to a challenge from another nation. It did not result because a mature society, having carefully assessed the state of its science and needs, decided that the time was right. Without the Soviet Union's first Sputnik, our mounting of the program to place a man on the moon ten years ago would have been at least as unlikely as our initiating a crash project today to unpollute all the Great Lakes.

The challenge to solve specific urban and environmental problems is here and more apparent to some citizens than the essentially invisible, though perturbing, first USSR satellite. But we are missing something that ten years ago led to a presidential proclamation to 'do it within the decade.'

Unfortunately, the Russians or Chinese are not about to 'threaten' us with a successful model city and thus cause us to accelerate our urban development. Similarly, the Moscow subway's existence has not driven the citizenry of Los Angeles to a crash development of a rapid transit system.

There is a different and important carry-over. A presidential call, even when backed by adequate funds from Congress, did not guarantee a successful lunar landing. The program had to be managed competently. No project involving many billions of dollars, hundreds of thousands of people, thousands of different contractors, numerous resource and facility creation problems, and countless overlapping steps can run well without good techniques of analysis, procedures for control and decision making, and a precise methodology for keeping track of everything and the relationship of one thing to another.

The 'systems approach,' as this way of going about the job usually is called now, is not magic. It is not even new. (Neither our pervasive telephone *system* nor our electric power distribution *system* came about by the random dropping from the skies of apparatus and people that just happened to work together well.)

What is new is the degree of development of specialized tools and skills of the professionals dealing with large-scale problems. The center of the systems approach is still, however, nothing but common sense and logic applied on a realistic basis.

The pattern of interrelationship and function of all elements to perform a job – people, machines, information flow, material – are conceived, specified, and analyzed, with the view of ending up with a coordinated, harmonious ensemble that meets the goals in an optimum way as judged by articulated criteria. The systems approach seeks to arrange the best trade-offs amongst the usually conflicting requirements.

It requires a knowledgeable and experienced team with specialists in all the technical and social facets involved. When attempting a difficult project, common sense must be embellished by lots of detailed knowledge, and an intellectual discipline for bringing that knowledge to bear on the problems pays off as against the usual random, piecemeal, emotional approach.

The systems approach will not yield answers to all problems. Even with an outstanding systems team, wrong judgment about an important factor or the lack of good information may cause the approach to fail. However, the systems approach is analysis and decision making based on a firm resolve to be complete, orderly, and logical. It is, at a minimum, a way to guard against chaos in the operations of our technological society.

To be sure, every 'social-engineering' problem – urban development, health care, rapid transit, education – possesses, to a degree very much beyond an intercontinental ballistic missile or communication satellite system, social factors that defy quantitative description. But the systems approach is still useful in such real-life problems. It is foolish to try to measure everything important in a problem and put it all down in a computer. It is equally foolish, however, to assume that nothing can be said with confidence about human problems, that we can get no guidance from scientific thought.

The systems approach will not eliminate disease, but if it is properly applied to the functioning of hospitals, it will cut costs and improve care. The systems approach to information flow in a large hospital complex has shown how to use better techniques for gathering and reporting test results, accounting information, and logistic supply data so as to minimize doctors' and nurses' precious time while providing more reliable and timely information.

Again, the systems approach will not eliminate automobile traffic congestion. However, when applied to the control of traffic flow, it increases the capacity of main arteries.

Computer is essential

Urban problems, despite their domination by social rather than technological factors, are not likely to be dealt with adequately without establishing the facts through much processing of statistics. For this, the computer is virtually an essential tool, there being much more to put down, remember, correlate and sort than the largest team of human minds conceivably can cope with. Categorizing and comparing the millions of facts required is easy for a computer, properly programmed.

Every social problem today – the aged, housing, poverty, transportation – has important technological aspects, either in the cause of the problem or as part of what ought to be done to help solve it. The expansion of industry and the agricultural-technology revolution, moving people from the country into the city rapidly before we were prepared for it, is the conspicuous example. In general, all technological factors can be analyzed with advantage by the systems approach, enabling the 'human' problem-solver then to concentrate better on the 'human' aspects of the problem that remain.

Some fear that the systems approach will lead to a society run by a computer. Of course, what they are really saying is that the systems approach will be improperly applied. They are not really arguing against the use of logic and objectivity. A *good* systems team has to include humanists as well as technologists, would not attempt to quantify the nonquantifiable, and would give full recognition to human factors even (and particularly) when they are known to be largely unknown. To omit considerations of man and look only at machines is not good systems management.

Man does have a choice as to what happens to his environment. However, to attain what he wants, he has to broaden more than his technological tools. He has to be willing to get into the decision-making position. Whether the project is a lunar landing or an improved system for rapid transit within his cities, he must *want* to do it. If he does, then he can use advantageously the management techniques of the first project on the second.

We can benefit by using the methodology that has landed men on the moon for our down-to-earth problems. But it will be a lot harder, specifically because it will require the committed participation of the citizen. It will also require, as did the manned lunar landing, decisive and bold decisions by the nation's leadership.

Further Reading

The most useful general book to have is the *General Systems Yearbook*, published annually by the Society for General Systems Research, Bedford, Massachusetts.

Readings 12 and 14 of this volume contain extensive bibliographies. Other useful specialist material is listed below

W. Buckley, *Sociology and Modern Systems Theory*, Prentice-Hall, 1966.

C. P. Bonini, *Simulation of Information and Decision in the Firm*, Markham, 1964.

M. L. Cadwaller, 'The cybernetic analysis of change in complex social organizations', *Amer. J. Soc.*, vol. 65, 1959–60, pp. 154–7.

R. M. Cyert, and J. G. March, *A Behavioural Theory of the Firm*, Prentice-Hall, 1963.

D. Easton, *A Systems Analysis of Political Life*, Wiley, 1958.

F. E. Emery (ed.), *Systems Thinking*, Penguin, 1969.

C. W. Golby, and G. Johns, *Attitude and Motivation: Committee of Inquiry on Small Firms*, Research Report no. 7, HMSO.

G. Goyder, *The Responsible Company*, Blackwell, 1961.

R. R. Grinker (ed.), *Towards a Unified Theory of Human Behaviour*, Basic Books, 1965.

A. D. Hall, and R. E. Hagen, 'Definition of a system', in *General Systems Yearbook*, Society for General Systems Research, 1959.

J. F. Hart, and S. Takasu (eds.), *Systems and Computer Source: University of Western Ontario Conference, 1965*, University of Toronto Press, 1966; Oxford University Press, 1968.

R. J. Hopeman, *Systems Analysis and Operations Management*, Prentice-Hall, 1969.

I. R. Hoos, *Systems Analysis in Social Policy*, Institute of Economic Affairs, 1968.

R. N. McKean, *Efficiency in Government Through Systems Analysis*, Wiley, 1958.

S. L. Optner, *Systems Analysis for Business Management*, Prentice-Hall, 1968.

H. A. Simon, 'Theories of decision making in economics and the behavioural sciences', *Amer. econ. Rev.*, vol. 49, 1959.

O. R. Young, 'A survey of general systems theory', *General Systems Yearbook*, vol. 9, Society for General Systems Research, 1964, pp. 61–80.

Acknowledgements

Permission to reproduce the following Readings in this volume is acknowledged to these sources:

1 The RAND Corporation
2 The RAND Corporation
3 The RAND Corporation
4 Human Sciences Research Inc.
6 Van Nostrand Heinhold Company
7 The RAND Corporation
9 McGraw-Hill Book Company
10 *Datamation*
11 Prentice-Hall Inc.
12 *Aerospace Management*
13 John Wiley & Sons Inc.
14 Stanford L. Optner
15 The Ronald Press
16 Delacorte Press
17 John Wiley & Sons Inc.
18 Dr William H. Mitchel
19 Simon Ramo and the *Los Angeles Times*

Author Index

Subject Index

MacNamara, Robert, 155, 156
Management
  control function, 298–9
  heuristic methods, 148
Management accounting,
  shortcomings, 295
Management information,
  accounting, 204–5
  data base, 207–10
  inventory control, 201–2
  management planning, 206–7
  merchandising, 205–6
  purchasing, 202–3
  sales, 200–201
  see also Information Systems,
    real-time
Management science, methods and
  objectives, 289–99
Management theory, developments,
  312
Managers, defensive tactics of, 275
Man – machine interaction, 92,
  101–2, 325
Methodology, 143
  axiomatic, 150
  normative, 150–51
  problem solving, 163–5
  systems analysis, 155
Military systems, 17, 19, 37, 121–4,
  139–40
  examples, 20
  force composition, 21–2
  management by tasks, 156
  variables in planning, 21–2
Minimax approach, 24–5
Models, 20, 46–7, 90
  business, 181
  classification by problem type,
    80–82
  design, 79
  dynamic, 26–7, 30
  mathematical, 85
  organizational, 301–2
  research, 79–81
Monte Carlo methods, 84–5, 136

NASA, 296
  planetary exploration, 222–6
Neumann, J. von, 304
Normative methodology and theory,
  150–51, 167

Objectives, 30
On-line systems, see Systems
Operations research, 17, 19–20,
  21–7, 37
Organizations
  and information, 175–7
  as systems, 300–302

Perception, 56–61
Performance, measures of, 290–91
Planning
  long-range, 143
  municipal, 260
  systems approach compared with
    inductive, 214
Planning, Programming, Budgeting
  System, (PPBS), 10, 156
Probability estimates, 17
Problem, definition, 183
Problem solving, 10–11, 17, 51,
  143–4, 162, 148–9, 167
Problems, recurring, mathematical
  models, 80–82
Process description, 88–9, 144
Purposive behaviour, 62–3

Ramo-Wooldridge Corporation,
  12–13
RAND Corporation, 12, 17,
  139–40, 155–61
  papers published by, 19, 37, 53, 121
Real-time systems, see Systems,
  on-line; Systems, real-time
Resource determination, 293–4
Risk-taking, 25–6

Self-regulation, see Cybernetics;
  Systems, stable
Sensitivity analysis, 224
'Single-thread' analysis, 253
Social organizations
  and cybernetics, 326
  and general systems organization,
    308–9
  and rational use of systems
    approach, 331–2
Stanford L. Optner and Associates,
  261
Stochastic estimates, 85
System
  acceptance, 283–5